ME

A Field and Garden Guide

Second Edition
Ivan Holliday

Published in Australia by Reed New Holland
an imprint of New Holland Publishers (Australia) Pty Ltd
Sydney • Auckland • London • Cape Town

14 Aquatic Drive Frenchs Forest NSW 2086 Australia
218 Lake Road Northcote Auckland New Zealand
86 Edgware Road London W2 2EA United Kingdom
80 McKenzie Street Cape Town 8001 South Africa

Copyright © 2004 in text, photographs and illustrations: Ivan Holliday
Copyright © 2004 New Holland Publishers (Australia) Pty Ltd

First published in 1989 by Hamlyn Australia
Reprinted in 1996
Revised edition published in 2004 by Reed New Holland

All rights reserved. No part of this publication may be reproduced, stored in a retrieval system or transmitted, in any form or by any means, electronic, mechanical, photocopying, recording or otherwise, without the prior written permission of the publishers and copyright holders.

The National Library of Australia Cataloguing-in-Publication entry for your forthcoming publication is as follows:

> Holliday, Ivan, 1926- .
> Melaleucas : a field and garden guide.
>
> 2nd ed.
> Includes index.
> ISBN 1 876334 98 3.
>
> 1. Melaleuca - Identification. I. Holliday, Ivan, 1926- A field guide to melaleucas. II. Title.
>
> 583.765

Publisher: Louise Egerton
Editor: Anne Savage
Designer: saso content & design
Production Manager: Linda Bottari
Reproduction: SC (Sang Choy) International Pte Ltd, Singapore
Printer: Kyodo Printing Co. Pte., Ltd.

Contents

Acknowledgements
5

Introduction
6

Cultivation
8

Floral Details
10

Descriptions of Species
Arranged Alphabetically by Scientific Name
12

Glossary
320

Bibliography
324

Index of Botanical and Common Names
326

Acknowledgements

My sincere thanks go to the following people who have assisted me in various ways.

- Lyn Craven of the National Herbarium, CSIRO, Canberra. Lyn headed the team responsible for revising the genus *Melaleuca*, and dealt so patiently with all my queries for confirmation of identity over many months. He and Brendan Lepschi included me on their field trip to WA in 1994 to collect and study melaleucas and I thank them both for their excellent company and assistance on this trip.

- Aileen Mehaffey, my sister (assisted on occasion by Margaret Blaber), deserves special credit for so diligently typing all the text, despite so many botanical names—a task which infiltrated her busy life over several months.

- Linda Gowing and Bruce Skinner, for scanning and printing my transparencies so that I could construct a mock-up of the book.

- Robert Smith, who drove me to photographic locations during periods of ill-health when I was unable to drive.

- Keith Pitman of Cockatoo Valley, SA, for his considerable help via his native plant plantation and for collecting and photographing specimens.

- Don Bellairs, Elizabeth George, Robert Smart, Alan Tinker and Coral Turley, all of WA, who assisted with local species on request.

- Don Overall of Victor Harbor, SA, for his cooperation with specimens at Nangawooka Reserve.

- Colin Cornford of Brisbane, for help with northern species.

- Bob Harwood of the Herbarium at Palmerston, NT, for assistance with photographs of tropical species by John Brock.

- Ian Bond, Lloyd Carman, Laurie Crooks, Murray Fagg, Alan Grünke, David Jones, Linda Niemann, John Simmons, Brenton Tucker and Geoff Watton for assisting with queries or photographs.

Introduction

A member of the very large family Myrtaceae, the genus *Melaleuca* is widespread in Australia, from the tropical north where paperbark swamps are common, through the arid inland, often in sandy or gravelly soils, to the wet, cold coastal areas of western Tasmania. Whilst Australia is the home of most *Melaleuca* species, a few tropical ones extend beyond Australia to the north, where they are found in places such as New Guinea, New Caledonia and Malesia. It was in fact from one of these species found outside Australia that the generic name was derived.

The genus was founded in the eighteenth century by Carolus Linnaeus and named from the Greek *melas*, meaning black, and *leucos*, white, apparently because of the tree's black trunk (probably charred by fire) and white-barked upper branches.

A team led by Lyn Craven of the National Herbarium, CSIRO, Canberra, has recently revised the genus, describing 255 species in Australia, some of which extend to Malesia and New Caledonia. *M. howeana*, from Australia's Lord Howe Island, is not included in this revision nor in this book. There are seven other species which occur only in New Caledonia.

The 255 species described for Australia, however, include 36 species which are currently *Callistemon*, and because this change has not yet been published or accepted, these 36 species have been omitted from this new edition. Furthermore, the revision team is examining some plants now under the *M. uncinata* umbrella, a study expected to add more species, three being included here (*M. concreta*, *M. hamata* and *M. osullivanii*).

This book then includes the 219 described *Melaleuca* species, the three named above in the *M. uncinata* study, 16 subspecies, 5 varieties and 10 garden forms and un-named species, a total of 253 different melaleucas. It totally revises and combines relevant information which was published in my earlier books on melaleucas.

Common names used for *Melaleuca* species include 'paperbark', referring to the papery bark common to a number of the larger tree species and sometimes used for lining hanging baskets; 'honey-myrtle'; and 'tea-tree', although this last name is more commonly reserved for the genus *Leptospermum*. Many melaleucas grow near water, often in swamps and estuaries, or along stream banks. In the arid inland they are often found growing in dry streams which are occasionally subjected to inundation; yet others grow happily in sandy soils with low to moderate rainfall, particularly in the sand heaths of Western Australia, where many species are prevalent. Others again grow along the exposed coastline, with different species to be found over most coastal habitats of the continent.

Melaleuca leaves yield several useful oils which are used commercially.

These include cineole and the valuable components of nerolidol (up to 90 per cent) and linalool (up to 30 per cent) used in the perfume industry. Oil from *M. bracteata* is a phenolic ether type, and that from *M. alternifolia* is a terpenic type, valued for its germicidal activity.

The well-known genus *Callistemon* (bottlebrushes) is closely related to *Melaleuca* but currently is separated from that genus by its flower structure. In *Melaleuca* species the stamens are united into bundles, while in *Callistemon* they are separate or free. Each individual flower in *Melaleuca* species contains 5 staminal bundles.

There are, however, a whole group of other myrtaceous genera in Australia that share this characteristic of having 5 staminal bundles in each flower. In most such genera, including *Melaleuca*, the inflorescence is a spike or rounded head of many individual flowers. The main means of identifying each genus is by the anther and ovary differences. These can be examined under a good magnifying glass, the ovary first being cut to display the ovules. In *Melaleuca* the ovary contains 3 chambers (loculi), and the anthers are attached at the middle (versatile). A sketch of *Melaleuca* floral details is included on page 10.

Cultivation

Melaleucas are excellent garden plants because of their general hardiness—nearly all withstand both wet feet and long dry spells, are suited to most soils, and can be cut and pruned to any desired shape or size without detriment. The only exceptions are a few species, mainly from the Western Australian sand heaths, that are a little tricky in some soils. Melaleucas do well in most parts of Australia, although many species from Western Australia's southwest and other southern parts of Australia can be difficult to establish in the more humid coastal areas, whilst most of the monsoonal species from the Top End swamps are not suited to areas south of Sydney.

Melaleucas may take the form of low dense bushes, suited to groundcover and excellent for living edges to lawns where the extra water required to maintain good grass cover might upset some other smaller natives; shrubs of all shapes and sizes suitable for ornament, screening or windbreak; or handsome specimen trees, such as *M. leucadendra*, *M. argentea*, *M. stypheliodes* and *M. linariifolia*. The 'Denmark' form of *M. microphylla* grows into a lovely, slender weeping tree that is useful for landscape work.

Many melaleucas are excellent plants for salty, low-lying, waterlogged soils, and indeed, some such as *M. halmaturorum* and *M. cuticularis* are possibly the only large plants which can be expected to succeed in soils of high salt content. Others are excellent plants for exposed coastal situations where they withstand wind-burn from salt-laden winds better than most plants—*M. nesophila*, *M. pentagona* var. *latifolia*, *M. leiopyxis* and *M. microphylla* are good examples in this category. Many species have handsome foliage year-round that responds to hard pruning and can be trimmed to any desired shape. Flowering is often prolific, although in many species it is of short duration. Those grown for their colourful flowers, once the bush has reached its desired size, are best cut back beyond the spent flowers each year immediately after flowering. This will ensure profuse flowering again the next year, and overcome the neglected look presented by a bush dominated by woody seed capsules.

The flowers come in virtually every colour except blue. They are white, cream, yellow, orange, red, pink, mauve, purple, green and shades in-between. In the reds there are subtle differences—the bright scarlet of *M. coccinea* and sometimes *M. macronychia*; the vermilion, or almost deep orange, of *M. lateritia*; the brick red to orange–red of *M. hypericifolia*; and the dusky red to almost maroon of *M. elliptica*. Pink to purplish pink is very common, but there are many variations including the delicate mauve–pink of some forms of the profusely flowering *M. pentagona* and the almost blush pink seen in *M. striata*, whilst an unnamed species I found near Esperance has dainty salmon pink flowers. Bright yellow is also fairly common, as exemplified by *M. laetifica* and *M. thapsina*.

White and cream flowers often cover the whole plant in species exhibiting these colours, such as *M. dissitiflora*, *M. linariifolia* and *M. viminea*. Green to yellow–green flowers can be found on *M. nervosa*, *M. diosmifolia*, *M. longistaminea* and *M. blaeriifolia*, and there is a form of *M. fulgens* with attractive apricot-coloured flowers.

Propagation of melaleucas is generally easy, both from seed and from cuttings. Seedlings, however, sometimes take from 7 to 20 years to flower, and may not come true to their parentage. Consequently, growers seeking a particular species they have seen are advised to grow it from cuttings to ensure early flowering and true form. A few semi-hardwood cuttings inserted in a tube of sharp sand and placed under a glass or plastic in semi-shade will usually be successful.

Floral Details

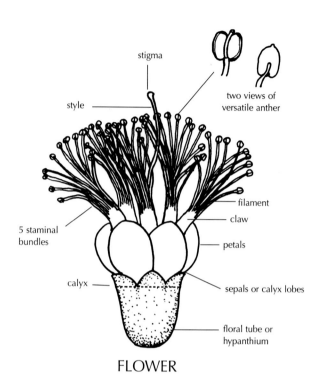

FLOWER

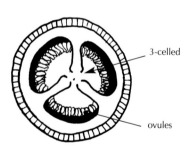

TRANSVERSE SECTION OF OVARY

SOME LEAF SHAPES

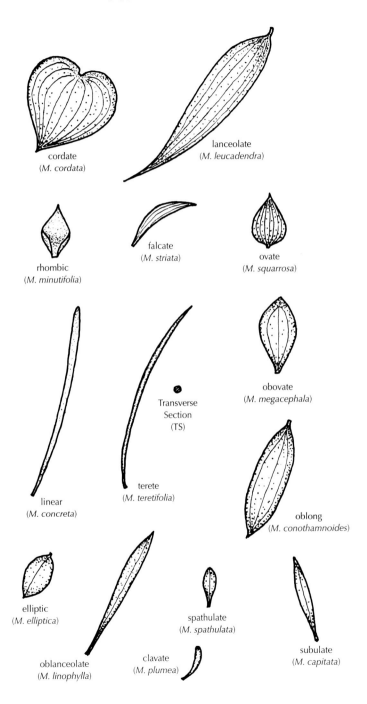

M. acuminata F. Muell. subsp. *acuminata*

DESCRIPTION ➤ Erect shrub, 1–3m, with many ascending branches.
LEAVES decussate, narrowly elliptic, 5–10mm long by 2–4mm wide, acuminate, with a small distinct petiole.
FLOWERS cream or white, occurring over long distances along the branches, in lateral clusters each on a very short axis; flowers subtended by several triangular bracts. Flowering season: spring.
FRUITS smooth, 3–5mm diameter, more or less globular; singly or several together, on a short stalk.

DISTRIBUTION ➤ Widespread in temperate parts of WA, SA, Vic., NSW.

DISTINGUISHING FEATURES ➤ Narrowly elliptic, acuminate, decussate leaves coupled with lateral creamy inflorescences along branches.

M. acuminata subsp. *websteri* (S. Moore) Barlow ex Craven, from Wubin to Wyalkatchem area of WA, differs mainly in its usually narrower leaves and shorter hypanthia and calyx lobes.

SIMILAR SPECIES ➤ *M. basicephala* Benth., a dwarf shrub from the freshwater swamps of the wet south of WA (Augusta to Northcliffe), has similar decussate leaves but pink or carmine flowerheads mainly on the secondary shoots. Rarely seen.

CULTIVATION ➤ Both subspecies of *M. acuminata* are very adaptable and easy to grow in most soils, acidic or alkaline, in dry temperate to temperate conditions. Frost hardy.
  M. basicephala is unknown to the author in cultivation.

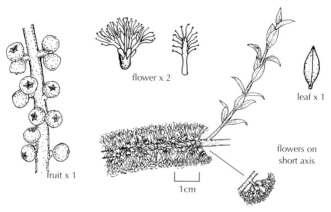

flower x 2

leaf x 1

flowers on short axis

fruit x 1

1cm

A large and adaptable screening shrub for most soils.

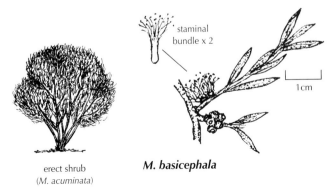

erect shrub
(*M. acuminata*)

M. basicephala

M. alsophila A. Cunn. ex Benth.

DESCRIPTION ➤ Dense shrub or tree to 15m, with fibrous or papery bark.
LEAVES feature wide variation in size and shape, even on the same plant, but commonly 10–50mm long (sometimes to 90mm), flat, 5–7 veined, spirally arranged, narrowly elliptic to narrowly obovate, the apex variable.
FLOWERS white or cream, occurring on small, dense, mostly lateral heads; flowers occur in dyads; hypanthium small, under 2mm long and wide; stamens 9–16 per bundle. Flowering season: dry season, March–October.
FRUITS in small, globular clusters, or a few together, each capsule cup- to barrel-shaped, about 2mm long and wide.

DISTRIBUTION ➤ WA, from Yampi Sound district south and west to 80 Mile Beach and Edgar Range districts, and from Cambridge Sound to Ord River district. Common in seasonally inundated areas near Broome.

DISTINGUISHING FEATURES ➤ Flat, 5–7 veined, acacia-like leaves and small, globular, white to cream flowerheads in dyads, occurring mainly in the leaf axils.

SIMILAR SPECIES ➤ *M. acacioides* (F. Muell.) Kuntze is closely related, and occurs from western Arnhem Land in the NT eastwards to Cape York Peninsula, Qld (also New Guinea), often on the landward side of mangroves and samphire in slightly saline soils. Amongst other features that differ, its inflorescences are in triads.

M. citrolens Barlow, another tropical paperbark, occurs from north-eastern NT to south-eastern part of Cape York Peninsula in Qld. It is of similar size with similar leaves to 90cm long, but the white to cream inflorescences differ by being in monads.

CULTIVATION ➤ None of these species is known in cultivation to the author, but should be suited to summer rainfall areas across the Top End. *M. alsophila* is best for the drier tropics.

A small tree suited to monsoonal planting.

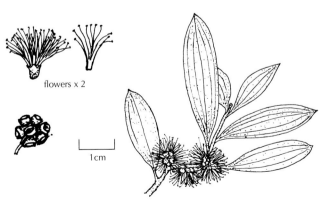

M. alternifolia (Maiden & Betche) Cheel

DESCRIPTION ➤ Small tree, to 7m, with a bushy crown and whitish papery bark.
LEAVES alternate, linear, 10–35mm long by about 1mm wide, smooth, recurved and soft; rich in oil, with the oil glands prominent.
FLOWERS in fluffy white masses, making a fine show over a short period; stamens pinnate on the claw. Flowering season: mostly spring to early summer.
FRUITS cylindrical, 2–3mm long and wide, usually sparsely spaced along branches.

DISTRIBUTION ➤ An inhabitant of wet places, mainly from Coffs Harbour in NSW north to Maryborough in Qld along the east coast, but recorded as far south as Stroud in NSW and inland to Stanthorpe district in Qld.

DISTINGUISHING FEATURES ➤ Massed heads of fluffy white flowers coupled with narrow, alternate leaves. *M. linariifolia* (p. 174) is similar in flower but differs in its wider, decussate leaves and flattish-spherical fruits.

CULTIVATION ➤ Grows well in a wide range of soils and climates, from Adelaide to North Queensland. Likes well-drained but moist soils and a sunny location; responds well to heavy pruning. Commercially cultivated for tea-tree oil, which is used as a germicide, for skin complaints and skin care.

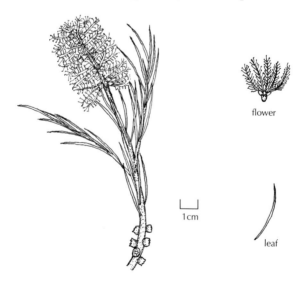

flower

1cm

leaf

Used commercially in the manufacture of tea-tree oil.

small, bushy-crowned tree

M. apodocephala Turcz. subsp. *apodocephala*

DESCRIPTION ➤ Low, bushy shrub, normally under 80cm.
LEAVES grey–green, glabrous, narrow, often wider towards the base, sessile, spirally arranged, pointed but not very prickly, mostly 6–12mm long, and slightly fleshy.
FLOWERS creamy white with yellow anthers; stalkless; in globular clusters along branches; hypanthium bell-shaped, glabrous; stamens mostly under 4mm long, 6–13 per bundle. Flowering season: usually October.
FRUITS about 5mm across, in globular or sub-globular clusters well embedded into corky branches.

DISTRIBUTION ➤ WA, from the Stirling Range east to the Truslove district, in sandy soils.

DISTINGUISHING FEATURES ➤ Crowded leaves, yellow, stalkless flowers in profuse lateral clusters, and corky branches.

M. apodocephala subsp. *calcicola* Barlow ex Craven, from near Scaddan in WA to western edge of Nullarbor Plain, differs in having more stamens per bundle (12–23) and a longer claw.

SIMILAR SPECIES ➤ *M. sculponeata* Barlow, an unrelated species, grows in the same area as *M. apodocephala* near Jerramungup. Very rare, tiny shrub to about 50cm high and wide, from Fitzgerald River district, with a disjunct occurrence near Lake King, featuring closely pressed, small, fleshy, decussate leaves, and white flowerheads from the laterals.

CULTIVATION ➤ None of these species have been cultivated to the author's knowledge, but could be used as low foreground shrubs in light, well-drained soils in dry temperate or temperate areas.

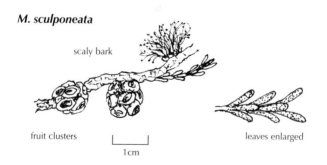

M. sculponeata — scaly bark, fruit clusters, 1cm, leaves enlarged

A low groundcover or foreground shrub.

M. apodocephala

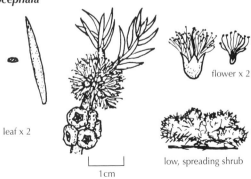

leaf x 2

1cm

flower x 2

low, spreading shrub

M. araucarioides
Barlow, in Quinn, Cowley, Barlow & Thiele

DESCRIPTION ➤ Small, rich green shrub, 0.5–1.5m high and 0.5–1m wide.

LEAVES fleshy, glabrous, oblong-linear or narrowly ovate, 2–6mm long by 1–2mm wide, obtuse, crowded and so regularly arranged around the smaller branches that they are almost cypress-like, or of herringbone appearance (ternate).

FLOWERS pale cream; in single heads or several together, terminal on the numerous upward-curving short branches, but lateral on the older wood; flowerheads 14–16mm wide in monads, usually only 3 stamens per bundle. Flowering season: spring.

FRUITS in small, tight clusters, the sepals persistent and petal-like.

DISTRIBUTION ➤ Restricted to Ongerup–Cape Richie–Jerramungup area in WA in woodland, shrubland and heath.

DISTINGUISHING FEATURES ➤ Small, fleshy leaves regularly arranged in herringbone fashion on smaller, normally upward-curving branches. This species is closely related to *M. blaeriifolia* (p. 28), from which it differs in its very distinctive ternate leaf arrangement and smaller, pale cream flowers with shorter styles and staminal filaments.

SIMILAR SPECIES ➤ *M. phoidophylla* Barlow ex Craven occurs in WA, from the Katanning district to Salmon Gums district. It is a shrub 0.2–6m tall, also featuring small leaves mostly ternately arranged, and white to cream flowerheads to 18mm wide, in monads. It has 7–11 stamens per bundle.

CULTIVATION ➤ The unusual foliage and numerous flowerheads of *M. araucarioides* at its best make this smallish shrub suitable for ornamental planting. Rarely seen in cultivation but is grown successfully in Adelaide in well-drained sandy soils. For winter-rainfall temperate areas.

small, low shrub

This small shrub features unusual foliage.

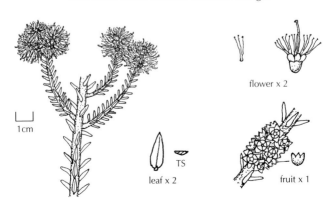

flower x 2

1cm

leaf x 2 TS

fruit x 1

M. argentea W.V. Fitzg.
Silver Cajuput, Silver-leaved Paperbark

DESCRIPTION ➤ Usually a decorative weeping tree, 8–20m, with papery bark and pale silvery green foliage.

LEAVES narrow-lanceolate, straight to falcate, 3–5 veined, 50–120mm long by 10–15mm wide, silvery grey–green, the young growth soft, silvery and silky-pubescent.

FLOWERS in cylindrical spikes 50–150mm long, terminal or in leaf axils, the axis growing on; pale greenish yellow to cream; individual flowers often in triads, well spaced along spike; stamens 3–7 per bundle. Flowering season: normally winter to spring.

FRUITS cylindrical or cupular, 3–4mm diameter, loosely spaced along the slender branches.

DISTRIBUTION ➤ Northern Australia, from the Kimberley of WA through Top End of NT to North Qld, favouring river banks.

DISTINGUISHING FEATURES ➤ Narrow-lanceolate, normally silvery leaves, weeping habit and papery bark.

SIMILAR SPECIES ➤ *M. fluviatilis* Barlow, from Cape York Peninsula–Cloncurry district south-east to Rockhampton in Qld, is a very similar paperbark, a tree or shrub to 30m, usually inhabiting stream banks or wet swamp margins. It differs in its different leaf hairs and longer, more richly coloured stamens (usually green).

CULTIVATION ➤ Handsome, weeping, silver-leaved tree for tropical to sub-tropical areas; often seen as an ornamental in Brisbane.

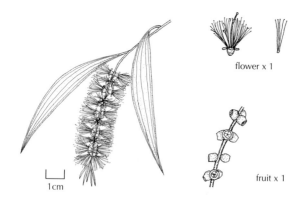

flower x 1

fruit x 1

1cm

A lovely weeping paperbark tree for the tropics and sub-tropics.

weeping tree

M. armillaris (Sol. ex Gaertn) Sm
subsp. *armillaris*
Bracelet Honey-myrtle

DESCRIPTION ➤ Small, dense-foliaged, deep green tree or large shrub, normally 5–8m, with rough, grey bark.
LEAVES usually 15–20mm long, spirally arranged and crowded, linear, very narrow (1mm or less), with a prominent hooked point, but not prickly.
FLOWERS in white, cylindrical, lateral spikes about 50mm long by 25mm wide, but variable. Flowering season: normally spring to early summer.
FRUITS about 5mm diameter, flattish, with enclosed valves, in clusters along branches.

DISTRIBUTION ➤ Coastal cliffs and river estuaries, from Manning River in NSW to eastern Vic., Tas. and Curtis Is. Very common in Mallacoota–Eden area.

DISTINGUISHING FEATURES ➤ Very narrow, crowded, hooked but not prickly, deep green leaves, giving a rich green, lacy or feathery look to the foliage; long spikes of flowerbuds surrounded by narrow, reddish bracts.

M. armillaris subsp. *akineta* Quinn, a rare subspecies occurring only in the Gawler Ranges of SA, differs in its shorter stamens and fewer flowers per spike.

CULTIVATION ➤ *M. armillaris* is one of the most commonly cultivated melaleucas, often used as a windbreak or for screening purposes. Very adaptable, but often abused when pruning is neglected. Some variable forms, possibly hybrids with species such as *M. hamulosa*, are in cultivation.

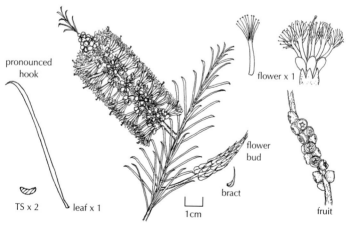

An extensively cultivated tree.

densely crowned small tree

M. armillaris subsp. *akineta*

M. aspalathoides Schauer

DESCRIPTION ➤ Usually a dwarf shrub under 60cm, but occasionally growing to 1m.

LEAVES narrow, terete, mostly about 10mm long, spirally arranged, crowded and silky-hairy, grey–green to grey. New growth leaves are longer (to 28mm); very soft and silky. Small tubercles visible beneath hairy surface.

FLOWERS mainly pseudoterminal or upper axillary, globular, magenta and very showy against greyish foliage; hypanthium woolly; calyx lobes sharply pointed; style extends well beyond stamens. Flowering season: usually late spring and summer.

FRUITS compressed-rounded to urceolate (older ones), to 7mm long by 5mm across.

DISTRIBUTION ➤ Walkaway area of WA south to Brookton–Tammin district.

DISTINGUISHING FEATURES ➤ Distinctive soft grey foliage and usually small stature; sharp-pointed, woolly calyx lobes when in flower.

CULTIVATION ➤ Excellent plant for rock gardens and small areas in light, well-drained loam or sand in dry temperate or temperate regions. Its permanent, soft grey foliage is an enhancing feature year-round. Adaptability to more humid or colder areas is unknown to the author.

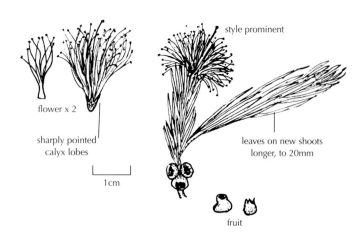

A lovely foliage shrub for the rock garden or border.

dwarf shrub

M. blaeriifolia Turcz.

DESCRIPTION ➤ Dense, intricately branched shrub, normally 1m or less and spreading, sometimes taller, to 2m.

LEAVES usually glabrous, spirally arranged, ovate to almost triangular, slightly petiolate, 2–6mm long by 1.1–2.5 mm broad.

FLOWERS greenish yellow, in globular to cylindrical heads, either at the ends of short side branchlets or in axils on older wood; hypanthium glabrous; stamens 3–5 per bundle. Flowering season: long period from winter to October.

FRUITS cylindrical, up to 6mm wide and long, either a few together, or forming a small spike, with sepaline teeth.

DISTRIBUTION ➤ Manjimup district of WA eastward to the Pallinup River, including the Porongurup Range, in sand or granite.

DISTINGUISHING FEATURES ➤ Small, ovate to triangular spreading leaves and small greenish yellow flowerheads.

CULTIVATION ➤ Grows best in sandy or light well-drained soils but adapts to non-alkaline heavier soils. Often forms a low groundcover shrub naturally, and can be trained to hug the ground. Successful in most situations from semi-dry to the humid east coast. Frost hardy.

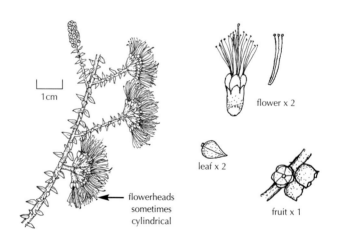

flowerheads sometimes cylindrical

flower x 2

leaf x 2

fruit x 1

A good foreground and groundcover shrub.

dense, spreading shrub

M. bracteata F. Muell.
Black Tea-tree

DESCRIPTION ➤ Bushy-foliaged small to medium tree, normally 5–8m but occasionally taller, with rough, dark grey bark.
LEAVES thin, linear-lanceolate to linear, 10–20mm long by 2–3mm broad, sessile, spirally arranged and crowded.
FLOWERS profuse, cream or white; loosely arranged in clustered, cylindrical or ovoid spikes 30–90mm long by about 15mm across. Flowering season: spring or early summer.
FRUITS ovoid to barrel-shaped, about 3mm diameter, sparsely arranged along branches.

DISTRIBUTION ➤ Widespread, from south-eastern NSW to Darwin in tropical NT, growing both near the coast and inland, usually along stream banks and in wet sites. Also in north-west SA to the Kimberley and Pilbara districts of WA.

DISTINGUISHING FEATURES ➤ Thin, linear-lanceolate, sessile leaves, rough, dark grey bark, and loosely arranged, normally cream flower spikes, often with a leaf growing at the base of the hypanthium. There are several garden cultivars, 'Revolution Gold' and 'Revolution Green' named for the colour of the foliage, and 'Golden Gem', a dwarf registered cultivar.

CULTIVATION ➤ Will grow almost anywhere, but does respond to extra water during very dry spells. 'Revolution Gold' is a popular small garden tree of attractive dense habit.

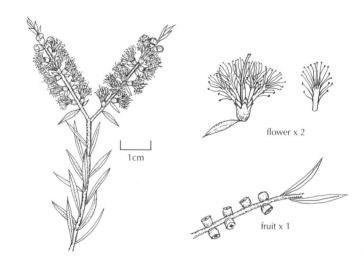

flower x 2

fruit x 1

Very adaptable small tree.

small to medium tree

M. bracteata 'Revolution Gold'

M. bracteosa Turcz.

DESCRIPTION ➤ Dwarf or small, spreading, dense shrub, normally about 0.5–1m high and wide, sometimes larger.

LEAVES tiny, spirally arranged, overlapping on younger tips but spaced out on more mature branches, 3–7mm long by about 1mm wide, glabrous, bright green, fleshy, with obtuse apex. Oil glands in rows.

FLOWERS profuse and showy at their best, normally bright cream, sometimes mauve–pink, in lateral rounded heads about 15mm across, new growth seldom growing on at maturity. Hypanthium glabrous; stamens 3–8 per bundle; flowers in singles (monads) on each head, and inflorescence axis hairy. Flowering season: August–November.

FRUITS small and cup-shaped, about 3–4mm by 3–4mm, in small clusters, or singly, calyx lobes reduced to sepaline teeth.

DISTRIBUTION ➤ WA, from Pingrup district south to Albany and east to Ravensthorpe, often under low trees or tall shrubs. Withstands occasional flooding.

DISTINGUISHING FEATURES ➤ Normally small stature, tiny, glabrous, fleshy, non-prickly leaves and lateral, usually cream flowerheads with flowers in monads, stamens 3–8 per bundle.

SIMILAR SPECIES ➤ *M. pomphostoma* Barlow, a rare species from Ravensthorpe–Hamersley River district of WA, is closely related but differs in its greenish yellow flowers with longer and more numerous (12–18) stamens. Flowering season: autumn–winter.

CULTIVATION ➤ Neither species is well known in cultivation but well-drained soils in a sunny location should achieve best results. Suited to temperate areas but soil adaptability is unknown.

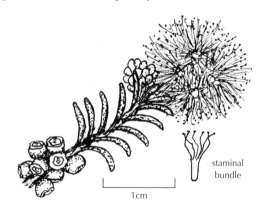

staminal bundle

1cm

A small, spreading shrub for temperate gardens.

small, spreading shrub

M. brevifolia Turcz.
Mallee Honey-myrtle

DESCRIPTION ➤ Small tree or bushy shrub, 2–5m, with tiny, dark leaves and rough bark.

LEAVES linear to narrowly linear-lanceolate, spirally arranged in whorls of 3, 4–10mm long by about 1mm broad, obtuse, with 2 rows of raised glands on the undersurface. Leaves on smaller branchlets straight, pointing upwards at very erect angle.

FLOWERS profuse, white or cream, in lateral clusters, spread over great lengths and irregularly along branches. Flowering season: spring.

FRUITS cup-shaped, 3–4mm long and wide, thick and corky.

DISTRIBUTION ➤ Widespread in SA saline sites that dry out in summer. Also in Vic. and WA, mainly along south coasts, in wide range of soil types.

DISTINGUISHING FEATURES ➤ Rough, dark bark, tiny leaves, normally with 2 rows of raised glands on undersurface, and profuse white to cream flowers clustered along the branches.

SIMILAR SPECIES ➤ *M. ordinifolia* Barlow, from Stirling Range and Hamersley River district of WA, is a small shrub usually under 1m, closely related but differing in its size and decussate leaf arrangement.

CULTIVATION ➤ *M. brevifolia* is a useful plant for harsh conditions such as the edges of saline clay pans and limestone soils, as well as for most other soils and conditions. Forms a good screening small tree or large shrub. For dry temperate or temperate regions.

M. ordinifolia is unknown in cultivation but is attractive in flower.

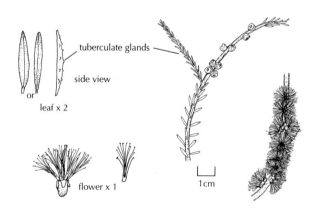

This is a useful plant for saline or limestone soils.

small, spreading tree
(*M. brevifolia*)

M. ordinifolia

crowded decussate leaves

1cm

M. bromelioides Barlow, in Barlow & Cowley

DESCRIPTION ➤ Stiff, conifer-like, glabrous shrub, to about 1–1.5m.
LEAVES crowded, pine-like ('*M. pinifolia*' would seem an apt name), blue–green, stiff and pungent, linear or linear-oblong, terete towards apex, usually 8–13mm long.
FLOWERS white, becoming reddish as they age, in small, terminal, capitate heads; buds red; each head about 15mm across. Flowering season: September–October.
FRUITS more or less cylindrical to barrel-shaped, with persistent triangular sepals, 4–5mm diameter by 5–6mm long, in small clusters or singly.

DISTRIBUTION ➤ Esperance region of WA, in a narrow zone from near Lake King to near Mt Heywood, mostly in mallee eucalypt association.

DISTINGUISHING FEATURES ➤ Very distinctive, pine-like, prickly foliage.

CULTIVATION ➤ Although the foliage may appeal to some, overall this shrub does not have much ornamental value for garden use. Will grow successfully in a range of soils, including some that become quite wet in winter.

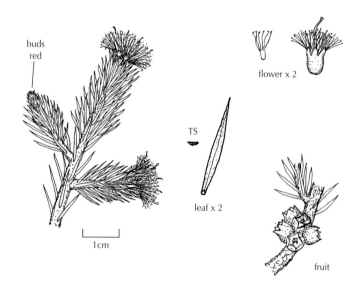

This smallish shrub has unusual pine-like foliage.

stiff, prickly shrub

M. brophyi Craven

DESCRIPTION ➤ Small to medium-sized erect shrub, seldom larger than 1m wide and high.
LEAVES spirally arranged, glabrescent, fleshy, warty, 5–16mm long, sub-terete, crowded, with prominent oil glands, apex obtuse or acute.
FLOWERS yellow, in small, globular, terminal heads about 14mm wide; rachis woolly; stamens 3–6 per bundle. Flowering season: mainly October.
FRUITS in small globular clusters; usually under 10mm diameter.

DISTRIBUTION ➤ WA, between Hyden, Southern Cross, Norseman and Esperance, usually in depressions which become wet in winter, and extending to Borden–Lake King and Ravensthorpe districts.

DISTINGUISHING FEATURES ➤ Small, fleshy, prominently oil-dotted, warty, sub-terete leaves, small terminal heads of yellow flowers and 'soccer-ball' fruiting clusters.

SIMILAR SPECIES ➤ *M. grieveana* Craven, from Cowcowing Lakes district to Narambeen–Parker Range district of WA, is very similar, differing in the leaf blades being clothed in pubescent to woolly-pubescent hairs.
  M. condylosa Craven, from Narambeen–Kondinin–Hyden district of WA, is also similar, distinguished by its knobby fruiting clusters, often a feature, and its smoother, longer leaves (9–32mm). Inflorescences are larger (to 20mm wide).

CULTIVATION ➤ None of these species is known to the author in cultivation, but they should succeed in dry to moderate, temperate conditions; they tolerate slightly saline, winter-wet soils.

M. brophyi

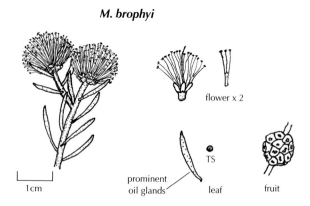

Attractive in flower — good for boggy soils.

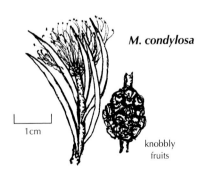

M. condylosa

1cm

knobbly fruits

small to medium erect shrubs

M. cajuputi Powell subsp. *cajuputi*

DESCRIPTION ➤ Shrub or more usually an erect, dense, paperbark tree to 35m tall.
LEAVES spirally arranged, mainly narrowly elliptic, 40–140mm long by 7–26mm wide, flat, glabrescent, but silky-hairy on the branchlets and silvery new growth.
FLOWERS white to greenish cream, in dense spikes 40–80mm long by 20–25mm wide, the axis growing on; hypanthium hairy or smooth; stamens 7–10 per bundle. Flowering season: March–November.
FRUITS cup-shaped, woody capsules, clustered loosely along branches.

DISTRIBUTION ➤ North-west WA and northern part of NT. Also Indonesia.

DISTINGUISHING FEATURES ➤ Medium to large, dense, erect, paperbark tree with large, narrowly elliptic leaves, silvery new growth and white or greenish flower spikes.

M. cajuputi subsp. *platyphylla* Barlow occurs in north-west Qld south to Cairns district, also in New Guinea and Torres Strait islands. It differs in its wider leaves (15–60mm), more stamens per bundle (8–13) and longer bundle claw.
   M. cajuputi subsp. *cumingiana* Turcz. Barlow occurs outside Australia in Malesia.

SIMILAR SPECIES ➤ *M. clarksonii* Barlow, occurring in central and northern districts of Cape York Peninsula, is a broad-leaved tree to 10m which differs in its hard, fissured bark, the only member of this group with this feature. The flowers and spikes are much smaller.

CULTIVATION ➤ All four trees should be easily grown in the wetter tropics and useful for shade or street trees. *M. cajuputi* subsp. *cajuputi*, extensively cultivated in Indonesia for the medicinal cajuput oil, also had many uses for the Aboriginal people—the wood for dugout canoes, the bark for a multitude of uses and the leaves for medicinal purposes.

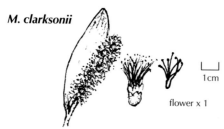

M. clarksonii

flower x 1

1cm

A tree for the tropics and medicinal uses.

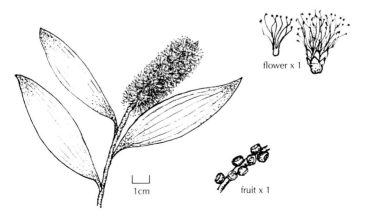

M. calothamnoides F. Muell.

DESCRIPTION ➤ Well-branched, erect shrub, usually to 1–4m high by 1.5m wide, with rough bark.
LEAVES spirally arranged, glabrous, fleshy, sub-terete to linear, mostly to 10–20mm long, obtuse and recurved.
FLOWERS in cylindrical spikes 40–50mm long and wide, on lateral shoots from old wood; normally pale green in centre, grading to red. Flowering season: usually late spring for some time, occasionally much earlier (July).
FRUITS in dense spikes 30–40mm long, each capsule 4–5mm across and tightly compressed.

DISTRIBUTION ➤ Murchison River area of WA.

DISTINGUISHING FEATURES ➤ Narrow, calothamnus-like foliage and normally reddish green, lateral flower spikes. Branches are very brittle.

CULTIVATION ➤ Sought after for its unusual red and green flowering spikes over a long period and its quite attractive foliage. This shrub has not proved reliable in cultivation other than in well-drained acidic soils, being a disappointment on limy and poorly drained soils. Regular pruning is recommended. Cuttings strike readily.

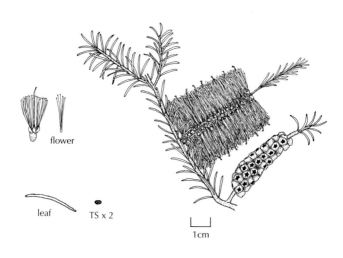

A brittle, long-flowering shrub for well-drained acidic soils.

erect, branching shrub

M. calycina R. Br., in Ait.

DESCRIPTION ➤ Stiff, normally erect shrub, to 2m but often less, with rough bark.

LEAVES opposite, stiff, glabrous, ovate to cordate, 6–10mm long by 5–7mm broad, sharply pointed, often recurved.

FLOWERS in leaf axils and terminally, singly or in clusters, each flower arising from a block of imbricate, silky-pubescent, brown bracts; petals and stamens white; hypanthium and lobes silky-pubescent. Flowering season: normally September.

FRUITS very distinctive, about 8mm long by 12–13mm across, with 5 long, curving, persistent sepals, star-like when viewed end-on.

DISTRIBUTION ➤ WA, from Stirling Range to Cape Arid and inland in places, mainly in low mallee and sandy heath.

DISTINGUISHING FEATURES ➤ Distinctive star-like fruits, stiff, pointed, ovate to cordate, opposite leaves, and prominent, imbricate brown bracts enclosing the flower buds.

SIMILAR SPECIES ➤ ***M. dempta*** (Barlow) Craven, occurring in Scadden–Gibson–Dalup River district of WA in dense scrub with sandy soils and on the edges of salty clay-pans, was first described in 1988 as *M. calycina* subsp. *dempta* Barlow. It differs in its fruits, which lack the persistent sepals, in its leaves, which have an obtuse apex, and in its smooth hypanthium. Given species status by Craven in 1999.

CULTIVATION ➤ *M. calycina* has occasionally been grown in southern gardens (Adelaide and Melbourne) in light or sandy soils. Its pure white flowers have a special appeal. Its extensive habitat range suggests that it could be grown in a wide variety of soils. Frost hardy.

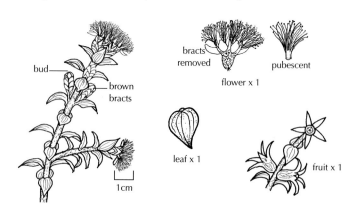

This smallish shrub is noted for its pure white flowers.

smallish, erect shrub

M. campanae Craven

DESCRIPTION ➤ Small, fairly nondescript, woody shrub, to 1m.
LEAVES glabrous, spirally arranged, mainly 20–40mm long by 3–5mm wide, narrowly obovate and acuminate; silky hairs closely pressed to branchlets.
FLOWERS mauve to mauve–pink in globular heads mostly 15–20mm across; calyx lobes usually absent, being replaced by an unbroken ring of tissue; hypanthium hairy. Flowering season: October–November.
FRUITS in sub-globose clusters, mostly pineapple-shaped and 8–12mm across.

DISTRIBUTION ➤ Small coastal band in Kalbarri–Geraldton district of WA, in sand and limestone, or on exposed coastal cliffs.

DISTINGUISHING FEATURES ➤ Mauve–pink flowerheads and sharp-pointed, narrowly obovate leaves to 40mm long. Look for the lobeless calyces.

SIMILAR SPECIES ➤ *M. eulobata* Craven, from the Shark Bay area of WA, is somewhat similar, differing in its grey–green, shorter, narrowly elliptic leaves and very short but distinctive calyx lobes, amongst other features.

CULTIVATION ➤ Neither species is known to the author in cultivation, but both could be useful plants for exposed coastal planting and on limy soils. Best for dry temperate to temperate regions.

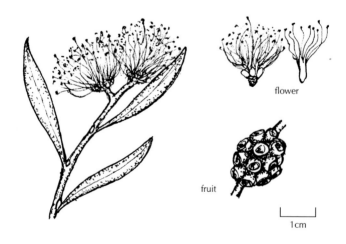

flower

fruit

1cm

M. campanae

M. eulobata

Both worth trying for coastal gardens and limy soils.

M. eulobata

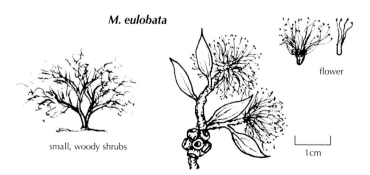

small, woody shrubs

flower

1cm

M. capitata Cheel.

DESCRIPTION ➤ Densely branched shrub, usually 1–2m high and wide.
LEAVES spirally arranged, glabrous, lanceolate, subulate or linear-elliptic, sharply pointed, 10–25mm long by about 2mm wide, with central nerve impressed on both surfaces; young growth villous.
FLOWERS pale yellow, in terminal, globular heads of 3–15 monads, with the axis growing on; hypanthium hairy. Flowering season: late spring to summer.
FRUITS about 6mm wide and long, barrel-shaped with deep sunken valves, in clusters.

DISTRIBUTION ➤ Coast and nearby mountains of southern NSW, along watercourses, from Bundanoon district to Braidwood. More common south of the Shoalhaven River.

DISTINGUISHING FEATURES ➤ Pale yellow, terminal, knob-like flowerheads, crowded, narrow, sharp-pointed leaves, and barrel-shaped fruits.

CULTIVATION ➤ Adapts to most soils and conditions where water is adequate, in full sun or semi-shade. Suitable for Sydney and probably Brisbane, as well as winter-rainfall temperate areas. Not known on Adelaide's limy soils, where it may suffer chlorosis. Frost hardy.

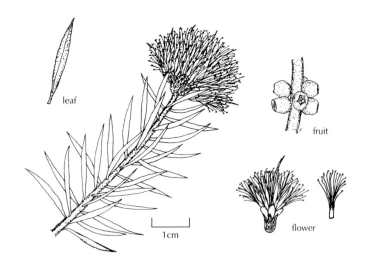

An attractive dense shrub for areas assured of rainfall.

branching shrub

M. cardiophylla F. Muell.

DESCRIPTION ➤ Dense, prickly, spreading shrub 1–3m high and wide, with many interwoven, tangly branches.
LEAVES alternate, glabrescent to hairy-peltate, ovate-cordate to ovate-lanceolate, undulate and recurved, striate, 4–10mm long by 2–6mm wide on average.
FLOWERS white, in sessile clusters over long distances along branches, stamens numerous and pinnate; both petals and sepals have hyaline margins; petals deciduous; hypanthium varies from green and smooth to woolly-hairy. Flowering season: variable, from August to summer.
FRUITS large, spherical, up to 10mm diameter, warty, with large sepaline teeth.

DISTRIBUTION ➤ Widespread in South-West Province of WA, from Exmouth district south to Perth district, and in the Pilbara, favouring limestone soils.

DISTINGUISHING FEATURES ➤ Prickly, ovate-cordate to ovate-lanceolate, peltate, striate leaves and white lateral flowers with pinnate stamens and hyaline margins to petals and sepals.

CULTIVATION ➤ Very tough, adaptable shrub for most soils and conditions, from semi-dry to temperate. Thrives on limestone where many other species founder. Tolerant of most frosts. Worth trying as frontline coastal plant, where it may be kept semi-prostrate by the winds.

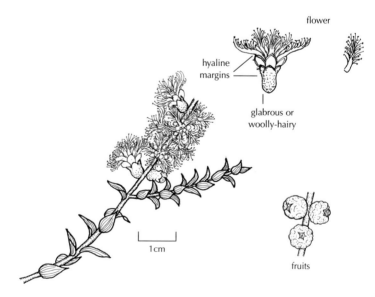

A tough shrub for most conditions including coastal gardens.

spreading shrub

M. cheelii C. White

DESCRIPTION ➤ Normally erect or spreading, a shrub or small to medium-sized tree to 8m high, with papery bark.
LEAVES decussate, elliptic, acute, 5–12mm long by 2–6mm wide, 3–5 veined, on a short petiole.
FLOWERS in loose terminal spikes about 40mm long by 20mm wide; white, but pink in bud; spikes comprised mostly of 2–10 monads; hypanthium hairy; stamens 8–18 per bundle. Flowering season: mid-spring.
FRUITS smooth, cup- or barrel-shaped, about 5mm long and wide, in loose clusters.

DISTRIBUTION ➤ Wide Bay district of Qld in heathy, sandy swamp.

DISTINGUISHING FEATURES ➤ Elliptic leaves in opposite pairs, cup-shaped or barrel-shaped fruits, loosely arranged, often in opposite pairs.

SIMILAR SPECIES ➤ *M. tortifolia* Byrnes, a closely related shrub found only in the Barren Mountain area west of Dorrigo in NSW, is distinguished by its usually larger, ovate, often twisted leaves and some floral differences.

   M. biconvexa Byrnes, from Port Macquarie south to Jervis Bay, NSW, is a shrub 3–8m tall with decussate, narrowly ovate or narrowly elliptic leaves 6–18mm long, distinguished by typically bird-winged transverse section. Flowers in cream spikes or heads occur in August–October.

CULTIVATION ➤ *M. cheelii* is grown successfully in Brisbane and is suited to sub-tropical areas of the east coast. Has succeeded in Adelaide on hard clay and red–brown earth, indicating it is very adaptable over a wide range of conditions. Author has no knowledge of the other two species in cultivation.

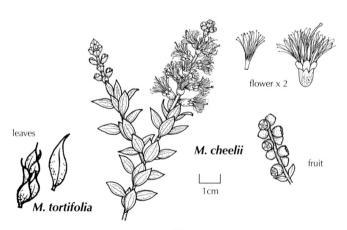

M. tortifolia

M. cheelii

An adaptable paperbark tree.

small to medium-sized tree

M. ciliosa Turcz.

DESCRIPTION ➤ Small, erect or straggly shrub, normally to about 1m high and wide, but variable.

LEAVES 4–12mm long, obovate to elliptic, sometimes narrowly so, spirally arranged, mostly recurved and ascending, shortly petiolate with visible, scattered oil glands; margins prominently ciliate.

FLOWERS in 3–15 triads on globular, profuse heads to 3 cm across, appearing yellow due to anthers and unopened buds, although stamens are creamy white; filaments age to pinkish red; glabrous calyx usually unlobed, the calyx an unbroken ring of tissue; if lobed, the lobes are without cilia; petals caducous; stamens 4–11 per bundle. Flowering season: usually October–November.

FRUITS may be of sub-globose type, with capsules tightly packed into an elongated cluster to 2cm long, or an irregular cluster where each capsule has separate identity. Capsules urn- or cup-shaped.

DISTRIBUTION ➤ Northern sand heaths of WA, from Badgingarra–Watheroo district to Mogumber district.

DISTINGUISHING FEATURES ➤ Obovate to elliptic leaves, although some other species have similar leaves, and bright yellow flowerheads with the calyx normally an unbroken ring of tissue.

SIMILAR SPECIES ➤ *M. lara* Craven, occurring in the Kalbarri district of WA, is similar. It is a shrub featuring generally smaller leaves, deciduous petals, more stamens per bundle (9–13), and yellow filaments which age to red, on smaller inflorescences.

CULTIVATION ➤ If kept tidy by regular pruning, *M. ciliosa* is an attractive small shrub for temperate to semi-dry areas, featuring showy flowers. Adapts well to a range of soil types and conditions, preferring well-drained soils and full sun for profuse flowering.

M. lara is unknown in cultivation but should perform like *M. ciliosa*.

M. lara

small shrubs

A very attractive yellow-flowering shrub.

M. ciliosa

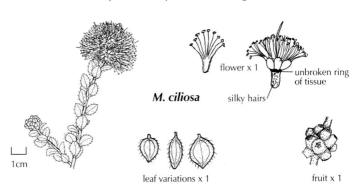

flower x 1 — unbroken ring of tissue, silky hairs

1cm

leaf variations x 1

fruit x 1

M. citrina Turcz.

DESCRIPTION ➤ Foliage of a pleasant, fresh green, and attractive flowers, are borne on a shrub of 1–3m, usually about 2m.
LEAVES flat, narrowly oblanceolate, 10–20mm long, obtuse to acute, glabrous, crowded and overlapping along branches.
FLOWERS sulphur yellow and very attractive, in oblong spikes about 20–25mm long, often in solitary pseudoterminal heads, individual flowers crowded tightly in spike. Flowering season: October–November.
FRUITS rounded to urceolate, 6–8mm across, compacted into clustered heads.

DISTRIBUTION ➤ Mainly in the Eyre district along the south coast of WA, but also the Stirling Range district near the sea.

DISTINGUISHING FEATURES ➤ Oblong, tightly packed, sulphur yellow flowerheads and crowded leaf arrangement.
   This species may not remain in *Melaleuca* because of its uncharacteristic seeds.

CULTIVATION ➤ Popular shrub in cultivation, successful in a range of soil types in full sun or semi-shade. Good drainage and regular pruning give best results but avoid limy soils. Requires a temperate climate with assured rainfall. Performance on the more humid east coast not known to the author. Strikes easily from cuttings.

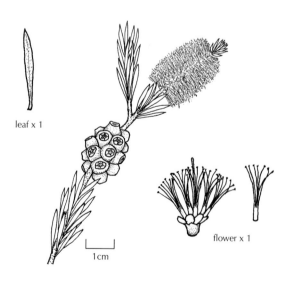

leaf x 1

1cm

flower x 1

A spreading shrub with very attractive flowers.

medium-sized shrub

M. coccinea A.S. George
Goldfields Bottlebrush

DESCRIPTION ➤ Medium-sized shrub, normally under 2m, with attractive foliage on numerous, slender, tangled branches.
LEAVES sessile, ovate to cordate, acute, reflexed, concave, and arranged in alternating opposite pairs (decussate); mostly 4–10mm long by 3–6mm.
FLOWERS brilliant red, in large, numerous, bottlebrush-like spikes 40–80mm long and to 50mm across, appearing on lateral shoots; buds covered by large cordate bracts; petals usually absent. Flowering season: spring and summer
FRUITS hairy capsules, tightly clustered to form cylindrical spike.

DISTRIBUTION ➤ Kalgoorlie–Norseman area of WA, mainly on granite outcrops, eastwards to the Chifley district.

DISTINGUISHING FEATURES ➤ Unmistakeable foliage, large red flower spikes, and flower buds covered by large cordate bracts.
 M. eximia (p. 106) and *M. penicula* (p. 106) are closely related and were previously subspecies of *M. coccinea*.

CULTIVATION ➤ Excellent garden shrub; tolerates most soils, including limestone. With extra summer water will flower well from November to March. Best suited to semi-arid and temperate areas. Frost hardy.

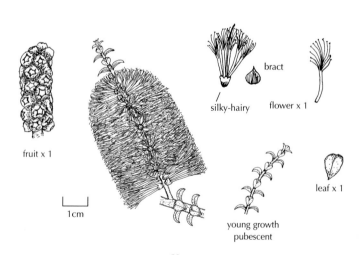

A long-flowering adaptable shrub.

medium-sized shrub

M. concinna Turcz.

DESCRIPTION ➤ Low, bushy, spreading shrub to 1m high.
LEAVES short, mainly 4–13mm long, terete, spirally arranged, narrowly clavate with sharp tip and short petiole 0.6mm or more long.
FLOWERS profuse, pink, in globular heads, both axially and terminally; hypanthium and floral axis hairy, hairs on outer hypanthium wall evenly distributed; floral bracts whitish and fluffy and petals papery, both persisting; usually only 3 stamens per bundle but numerous flowers on each head. Flowering season: mainly October–November.
FRUITS small, forming 'soccer-ball' cluster, 6–7mm diameter.

DISTRIBUTION ➤ From Stirling Range to Jerramungup area of WA.

DISTINGUISHING FEATURES ➤ Fluffy, white, persistent floral bracts in association with 'soccer-ball' fruits. Cf. *M. plumea*, which has similar features but loosely arranged or 'peg-like' fruits (see p. 219).

CULTIVATION ➤ Attractive small, bushy, profusely flowering shrub suited to edging a garden bed, provided it is pruned each year after flowering once it has reached desired size. Not often cultivated, but has proved adaptable in Adelaide in full sun in deep sand and quite heavy clay soils, and should adapt to most semi-dry to moist temperate areas of winter rainfall.

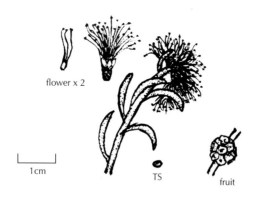

An excellent profusely flowering small shrub.

low, bushy shrub

M. concreta F. Muell.

DESCRIPTION ➤ Erect, small to medium-sized shrub 0.6–2m high and wide.
LEAVES spirally arranged, linear or very narrowly lanceolate, with straight or uncinate mucro, widely spaced, glabrous to glabrescent, new growth silky-hairy; 30–80mm long by 2–4mm wide, narrowly elliptic in transverse section, with oil glands in 2 rows.
FLOWERS profuse, cream to yellow, in axillary and terminal, globular heads, normally 10–15mm across. Flowering season: usually mid- to late spring.
FRUITS in compact, ovoid or round 'soccer-ball' clusters; individual capsules 2–4mm across.

DISTRIBUTION ➤ From northern Irwin district of WA, north of the Murchison River, often in alkaline sand near the coast.

DISTINGUISHING FEATURES ➤ Narrow, flattish, widely spaced leaves (transverse section narrowly elliptic), with a sharp but not prickly mucro, oil glands in 2 rows and silky young; globular, axillary and terminal, cream to yellow flowerheads; fruit of 'soccer-ball' type.

This species has been included at times under the various forms of *M. uncinata* (p. 300)

CULTIVATION ➤ Grows well in temperate to moderately dry areas of winter rainfall in sand or clay, which may be alkaline or acidic. Showy in flower. Could be tried in wind-affected coastal situations.

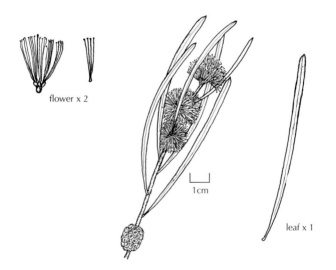

flower x 2

1cm

leaf x 1

An adaptable shrub for temperate areas including coastal gardens.

erect shrub

M. conothamnoides C. Gardner

DESCRIPTION ➤ Many-branched, low, woody shrub, usually 0.5–1m high and across.

LEAVES spirally arranged, oblong to oblong-lanceolate or elliptic, smooth and rigid, 25–40mm long by 6–8mm wide, 3–5 nerved, with prominent oil glands; young leaves ciliate.

FLOWERS prominent, in globular or ovoid heads of deep pink–purple, fading to white as they age, making a fine show over whole bush; base of hypanthium covered with soft white hairs; large, greenish black bracts surround unopened buds. Flowering season: usually September–November.

FRUITS formed into tight, globular or ovoid cluster of woody capsules.

DISTRIBUTION ➤ Inland, favouring sandy or gravelly soils between Morawa and Tammin in drier wheatbelt of south-west WA.

DISTINGUISHING FEATURES ➤ Large, rigid, oblong to oblong-lanceolate or elliptic leaves, usually with prominent oil glands, prominent pink–purple flowerheads and large, greenish black bracts covering unopened buds.

SIMILAR SPECIES ➤ *M. barlowii* Craven, occurring in Mullewa–Morawa district of WA, is a similar but more erect shrub, the leaves mainly narrower with different venation, and fruiting clusters less rounded. Flowers November–December.

CULTIVATION ➤ *M. conothamnoides* is an attractive garden shrub best suited to light or well-drained soils in dry to moderate, temperate areas; difficult in summer rainfall areas such as the east coast.

The author has no knowledge of *M. barlowii* in cultivation.

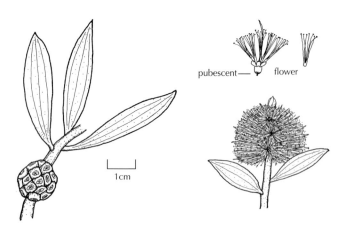

In flower, a very showy garden shrub.

M. barlowii

low, spreading shrub
(*M. conothamnoides*)

M. barlowii

flower

small, open shrub

M. cordata Turcz.

DESCRIPTION ➤ Smallish, rather rigid shrub, seldom exceeding 1m high and wide, with mainly erect branches.
LEAVES spirally arranged, sub-sessile to shortly petiolate, cordate, up to 30mm long and wide, glaucous, mainly 5–8 nerved, with a small mucro or acuminate apex.
FLOWERS deep pink to purplish red, in ovoid or globular heads 30mm or more across, terminal, or thickly clustered towards ends of branches. Flowering season: long period from late spring to mid-summer.
FRUITS are capsules about 4mm by 4mm, forming ovoid clusters.

DISTRIBUTION ➤ WA, from Geraldton–Mullewa districts south to Lake Grace–Lake King districts and eastwards to Coolgardie district, in variable habitats.

DISTINGUISHING FEATURES ➤ Prominent, glabrous, heart-shaped leaves to 30mm long and wide.

SIMILAR SPECIES ➤ *M. orbicularis* Craven, occurring from Coorow to Wongan Hills and Wyalkatchem in WA, is a small, erect shrub featuring similar but much smaller heart-shaped leaves, with a rounded apex.

CULTIVATION ➤ *M. cordata* makes an attractive, somewhat unusual garden shrub; it grows successfully in sand or clay in temperate areas of low to moderate winter rainfall. Showy flowers over a long period (mainly early summer) add to its appeal.
  M. orbicularis is unknown by the author in cultivation.

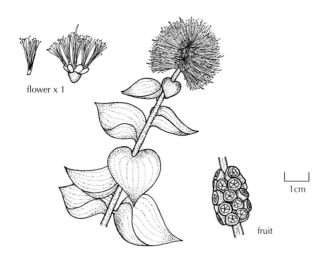

flower x 1

1cm

fruit

This species flowers for a long time through summer.

M. orbicularis

small, rigid shrubs

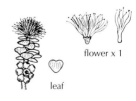

flower x 1

leaf

***M. orbicularis* x 0.5**

M. cornucopiae Byrnes

DESCRIPTION ➤ Slender shrub, 1–4m, with papery bark.
LEAVES spirally arranged, 40–105mm long by 4–15mm wide, stiff, narrowly elliptic to narrowly obovate, silky-hairy becoming smooth, with longish stalk and distinct longitudinal veins.
FLOWERS white, in spikes of 10–50 triads, each spike up to 50mm long by 15mm wide; hypanthium hairy; stamens 5–6 per bundle. Flowers open from bottom of spike in the distinctive manner seen in photograph. Flowering season: December–April.
FRUITS woody capsules 3–4mm long and wide, packed into cylindrical spikes; seed is shed annually.

DISTRIBUTION ➤ Western Arnhem Land in NT, including Kakadu National Park. It is endemic to the sandstone escarpment of the region.

DISTINGUISHING FEATURES ➤ Slender, shrubby habit, stiff leaves, and white flowers opening from base of spike.

TWO OTHER TROPICAL SPECIES ➤ ***M. arcana*** S.T. Blake, shrub 0.5–15m high, occurring from the tip of Cape York Peninsula to near Cooktown in Qld, is common in moist swales between sand dunes. Flowerheads are white, the leaves somewhat similar to *M. cornucopiae* but shorter and wider. Seed is shed annually. Flowering may occur at any time.

M. stipitata Craven, also from the Top End, and discovered only in 1991, is a shrub to 4m, found only at one location, below Bukbukluk Lookout on the Kakadu Highway in Kakadu National Park. It features linear to various-shaped, narrow leaves, 18–75mm by 1–5mm and white to cream loose flower spikes of 3–10 triads to 15mm wide in December.

CULTIVATION ➤ Cultivation unknown by the author but all should grow well in tropical conditions and on the east coast with summer rainfall.

staminal bundle x 1

1cm

M. arcana

An unusual shrub for the tropics.

slender shrub

M. coronicarpa D.A. Herbert

DESCRIPTION ➤ Spreading, prickly shrub to 2m, with flaking grey and white bark.
LEAVES ovate to very narrowly ovate, sessile, mostly 6–10mm long by about 2–6mm wide, spirally arranged and sharp pointed.
FLOWERS white, occurring laterally along branches in monads, forming long clusters; hypanthium hairy; stamens 10–22 per bundle. Flowering season: mainly spring.
FRUITS mostly cup-shaped, calyx lobes persisting to form obvious teeth on fruit.

DISTRIBUTION ➤ Widespread in WA, from near Binnu south to Ongerup district and inland to Koorda and Grass Patch districts.

DISTINGUISHING FEATURES ➤ Prickly, ovate leaves, lateral clusters of white flowers in monads and fruits with prominent teeth.

SIMILAR SPECIES ➤ *M. undulata* Benth., from Stirling Range–Broomehill district of WA eastwards to Israelite Bay, is a very similar shrub but can be readily separated by the sepaline teeth on the fruit, and wavier leaves with distinct oil glands.

M. delta Craven, a shrub occurring disjunctly in Kalbarri, Jurien and Wongan Hills districts of WA, features much shorter styles than *M. coronicarpa*, 2–4mm cf. 7–11mm, and white filaments with tiny rounded protuberances on the surface. Leaves are more elliptic.

CULTIVATION ➤ Rarely cultivated, but all three species could be used as dense, prickly barriers for controlling foot traffic. All very adaptable to a wide range of soil types and conditions in winter rainfall areas.

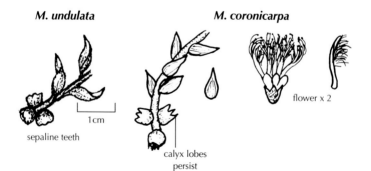

M. undulata — sepaline teeth — 1cm

M. coronicarpa — calyx lobes persist — flower x 2

This prickly shrub has attractive flowers.

prickly, spreading shrub

M. croxfordiae Craven

DESCRIPTION ➤ Leafy paperbark tree, normally bushy-crowned, to 8m at its best, but a low shrub in exposed coastal situations.
LEAVES thin and flat, linear or linear-elliptic, 20–60mm long by 2–4mm broad, pointed but not prickly, the 5 veins not visible; young growth softly ciliate.
FLOWERS white to cream, in small globular heads, 15–20mm across, and not numerous; hypanthium slightly hairy; stamens 5–8 per bundle. Flowering season: normally November–December.
FRUITS form globular, tightly packed heads, 12–15mm in diameter.

DISTRIBUTION ➤ Manjimup–Albany district of WA, abutting streams, in winter-wet swamps, in shrubland and occasionally in exposed coastal locations.

DISTINGUISHING FEATURES ➤ Long, linear, thin, flat, soft leaves, papery bark and small, globular flowerheads produced sparsely often in early summer.

CULTIVATION ➤ Despite its normally wet habitat this species is a success in the dry clay soils of the Adelaide region, suggesting it is an adaptable species and may be suited to a wide range of temperate conditions.

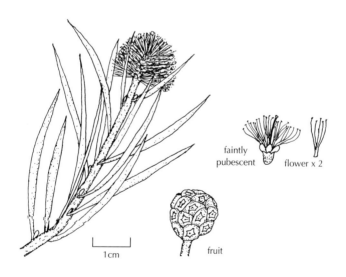

An attractive small paperbark tree.

leafy tree

M. ctenoides Quinn

DESCRIPTION ➤ Loose, spreading shrub, 1–3m, with rather brittle branches.

LEAVES smooth, with pronounced pustules on the older leaves, mostly 20–30mm long, semi-terete to terete, spirally arranged and curving upwards from one side of branch like the teeth of a loose comb (hence 'ctenoides' from ctenos, Greek for 'comb').

FLOWERS profuse and laterally arranged in showy, violet spikes 20–30mm long by 25mm across, the axis growing on; flowers loosely arranged in monads; hypanthium glabrous; petals white to mauve. Flowering season: mid- to late spring.

FRUITS barrel-shaped, 4–5mm long by 3–4mm across, in spikes or loose clusters.

DISTRIBUTION ➤ Mainly across a narrow band of south-west WA in transitional zone from moderate to low rainfall, from inland of Morawa to Narambeen–Hyden district, in mallee or heath.

DISTINGUISHING FEATURES ➤ Comb-like leaf arrangement, sub-terete to terete leaves and showy violet or mauve flowers.

This species is very close to the more widespread *M. laxiflora* (p. 164). In some places they grow together but *M. ctenoides* is easily separated by the form of its leaves and their arrangement.

CULTIVATION ➤ *M. ctenoides* is rarely cultivated, but is known to have succeeded in SA on well-drained, slightly alkaline loam. The more commonly cultivated *M. laxiflora* is very adaptable and *M. ctenoides* would probably perform likewise. Limy soils cause the foliage to become chlorotic, and should be avoided.

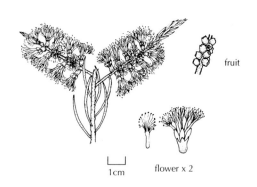

fruit

1cm   flower x 2

This medium-sized shrub features showy flowers in spring.

loose, spreading shrub

M. cucullata Turcz.

DESCRIPTION ➤ Normally a large, dense shrub, 2–4m high and wide, often with slender, arching branches projecting from the main frame.
LEAVES usually decussate, ovate-triangular, thick and fleshy, 2–4mm long and closely alternate, sometimes very close in appearance to *M. huegelii*.
FLOWERS in showy ovoid heads, 15–18mm long by about 15mm across; normally vivid white, profuse; hypanthium glabrous; 5–9 stamens per bundle. Flowering season: September–November.
FRUITS form ovoid spike, usually about 17–20mm long.

DISTRIBUTION ➤ Eyre district of WA, where it is fairly common from Salmon Gums to Esperance, and extending from Lake Grace–Stirling Range districts east to Israelite Bay, on a variety of soils and habitats.

DISTINGUISHING FEATURES ➤ Thick, ovate-triangular leaves, normally in opposite pairs, and showy, white, ovoid flower spikes. Although in some forms the leaves resemble those of *M. huegelli* (p. 140), the flowerheads are quite different and clearly separate the two species.

CULTIVATION ➤ The author is unfamiliar with this species in cultivation but it should form a large, attractive flowering shrub with judicious shaping. Habitat range suggests it would succeed on most soils in a semi-dry to temperate climate.

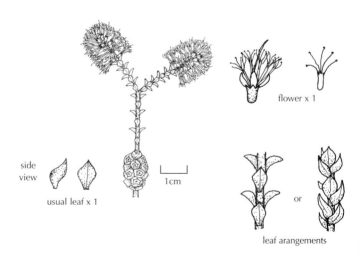

A large, attractive shrub for temperate regions.

tall, woody shrub

M. cuticularis Labill.
Saltwater Paperbark

DESCRIPTION ➤ Large shrub or eventually a tree, 6–12m high, branches tortuous, particularly on young plants, bark papery and white.
LEAVES decussate, glabrous, linear to linear-oblong, on average 10mm by 2mm, often markedly recurved.
FLOWERS white or cream, in single or multiple monads (1–3) at ends of branches, and surrounded by imbricate brown bracts. Flowering season: normally September–October.
FRUITS rough and woody, up to 10mm long, claw-like due to the 5 long persistent sepals, star-shaped when viewed end-on.

DISTRIBUTION ➤ Common in south-west of WA, from Perth to Israelite Bay, often in saline depressions and swamps. Also a rare occurrence on Kangaroo Island, SA.

DISTINGUISHING FEATURES ➤ Fruits with long, persistent, claw-like sepals, star-shaped in end-view; narrow, evenly decussate leaves; and very white, papery bark.

SIMILAR SPECIES ➤ Several poorly known but similar shrub species occur in the *M. cuticularis* complex, including:
M. haplantha Barlow, from the New Norcia district of WA eastwards to Mukinbudin and south of Esperance, is closely related but differs in the corky fruits lacking woody, sepaline teeth, narrower, sharper leaves and its shrubby habit, amongst other differences.
   M. sciotostyla Barlow, a rare species from near Wongan Hills in WA, is also of shrubby habit (1–1.5m high), but has narrower leaves and less stamens per bundle (12–17) as well as up to 4 flowers per inflorescence.

CULTIVATION ➤ *M. cuticularis* is an excellent small tree for saline, swampy conditions in most soil types and will tolerate moderate coastal exposure. Suited to winter-rainfall temperate regions.
M. haplantha should adapt to similar conditions. *M. sciotostyla* is unknown to the author in cultivation.

An excellent tree for wet saline soils.

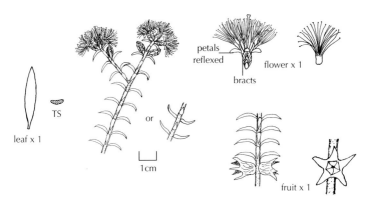

M. dealbata S.T. Blake

DESCRIPTION ➤ Medium to large tree to 25m, with leafy canopy, whitish or greyish, layered, papery bark and blue–grey, dull foliage.
LEAVES quite large, elliptic to narrowly obovate, apex normally acute, 50–120mm long by 10–30mm wide, spirally arranged and scattered, with prominent longitudinal veins; new growth silky-grey.
FLOWERS cream, in triads (7–28) loosely arranged on narrow spikes up to 120mm long by 25mm wide; stamens less than 10mm long, and 5–8 per bundle. Flowering season: mainly spring, with spasmodic flowers at other times.
FRUITS cupular or barrel-shaped, 3–4mm by 3–4mm; seeds shed annually.

DISTRIBUTION ➤ Swampy areas near the coast and inland, from north of Maryborough in Qld through north Qld to the Top End of NT and into Broome–Derby and Weaber Range districts of WA.

DISTINGUISHING FEATURES ➤ Large tree with dull blue–grey to grey–green leaves, long, narrow, loosely arranged cream flower spikes, the flowers in triads with short stamens.

OTHER TROPICAL SPECIES ➤ Three other species from the tropics are:
M. saligna Schauer, in Walpers, extending from Torres Strait islands to near Cooktown in Qld, is a small, pendulous, paperbark tree to 10m, with light green leaves to 120mm long, and white to greenish yellow flowers of 5–15 triads in terminal spikes or axillary capitate heads.

M. sericea Byrnes, from the eastern Kimberley to western parts of NT, is a smaller paperbark, 3–7m, with silky-hairy leaves to 65mm long, and whitish flowers occurring on spikes or capitate heads of 2–9 triads.

M. stenostachya S.T. Blake, from Borrooloola district of NT to Cape York Peninsula, is closely related to *M. dealbata*. It is a shrub or tree 2–15m tall, with white or cream flowers in loose spikes of 6–12 triads, distinguished by each triad being at least 1 hypanthium distance apart.

CULTIVATION ➤ All these species are suited to the tropics and warmer sub-tropics, especially in wet locations. The pendulous habit of *M. saligna* is particularly ornamental.

Here's one for the tropics with handsome bluish foliage.

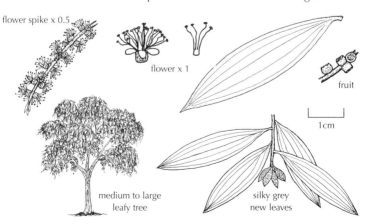

flower spike x 0.5

flower x 1

fruit

1cm

medium to large leafy tree

silky grey new leaves

M. decora (Salisb.) Britten

DESCRIPTION ➤ Small, rather decorative, slender tree, usually 5–10m, with brownish white, papery bark.

LEAVES spirally arranged, glabrescent, linear to linear-lanceolate, acute, normally 10–20mm long by 1.5–2mm wide, one side channelled by depressed midvein.

FLOWERS cream or white, sweetly perfumed, in cylindrical to ovoid spikes 20–50mm long by 20mm wide; hypanthia mostly smooth; stamens 20–40 per bundle. Flowering season: normally summer.

FRUITS more or less spherical, about 3mm diameter, in sparse clusters along branches.

DISTRIBUTION ➤ From Brisbane region in Qld south to Nowra in NSW, favouring coastal swamps. Common around Sydney.

DISTINGUISHING FEATURES ➤ Small paperbark tree with linear to linear-lanceolate leaves, spherical fruits with enclosed valves. Profuse, sweetly perfumed, cream flower spikes.

CULTIVATION ➤ Particularly decorative small tree in full flower; grows well in temperate and subtropical areas where water is adequate. Excellent for wet, swampy sites. Frost hardy.

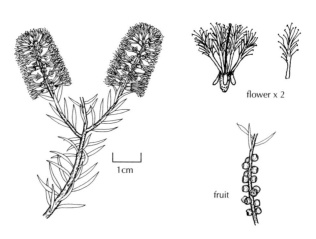

flower x 2

1 cm

fruit

A decorative paperbark for wet situations.

small tree

M. decussata R. Br., in Ait.
Totem Poles, Cross-leaved Honey-myrtle

DESCRIPTION ➤ Many-branched, normally erect shrub, 2–3m tall; a variety of forms found in cultivation.

LEAVES decussate, sessile, glabrous, narrowly oblanceolate to almost linear, obtuse to acute, 5–15mm long by 1–4mm broad, with prominent glands on undersurface.

FLOWERS violet or purplish to pink, in small, cylindrical spikes, 20–30mm long by 10mm across, subtended by deciduous, narrowly triangular bracts as long as leaves. Flowering season: mostly early summer.

FRUITS distinctive, with capsules fused together at base, becoming immersed in inflated rachis, a feature shared with the similar *M. gibbosa*.

DISTRIBUTION ➤ Common in wetter coastal regions of SA, extending from the south-east to Eyre Peninsula, and also found in western Victoria.

DISTINGUISHING FEATURES ➤ Fruiting capsules fused in opposite pairs, narrowly oblanceolate; decussate, sessile leaves. The similar *M. gibbosa* (p. 122) has shorter, more rounded, ovate to obovate leaves, but the occurrence of intermediate leaf forms makes the 2 species difficult to separate at times.

CULTIVATION ➤ Very adaptable shrub, grown successfully in all capital cities except Darwin on a variety of soils. Prune regularly to avoid a straggly, woody bush. Frost hardy.

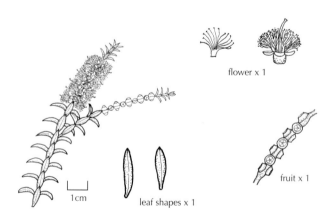

flower x 1

fruit x 1

1cm

leaf shapes x 1

An easily grown shrub for most garden situations.

erect shrub

M. densa R. Br., in Ait.

DESCRIPTION ➤ Very variable bushy shrub, normally 1–2m high and wide, the branches divaricate and the greyish bark fibrous.
LEAVES in whorls of 3 or 4 around branches, broadly ovate, or elliptic, recurved, sub-sessile, 3–10mm long; leaves on soft, young shoots often larger than normal and overlapping; older leaves may turn red during the cold months.
FLOWERS profuse, cream to pale lemon, in ovoid-oblong heads 15–25mm long; stamens 3–6 per bundle; buds surrounded by deciduous bracts. Flowering season: normally August–September
FRUITS cupular, with 5 persistent, rounded, sepaline teeth and slightly raised valves.

DISTRIBUTION ➤ WA, from Augusta to Albany–Stirling Range district.

DISTINGUISHING FEATURES ➤ Ovate leaves, usually in whorls around the branches, and soft young growth with overlapping, larger leaves.

SIMILAR SPECIES ➤ *M. pritzelii* (Domin.) Barlow, from Stirling Range and surrounds to Bremer Bay of WA, was once a named variety of *M. densa*. However, the specimen examined by the author, with its small ovate leaves and greenish cream inflorescences in the laterals, appeared much more like *M. blaeriifolia* (p. 28).

CULTIVATION ➤ *M. densa* has been grown successfully in Adelaide's alkaline clay soils but is not often seen in cultivation. Should adapt to most soils in temperate areas where annual rainfall exceeds 400mm. Regular pruning is necessary to keep the plants shapely.
  M. pritzelii is unknown to the author in cultivation.

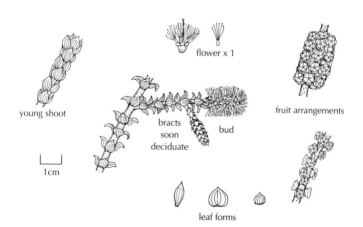

young shoot
flower x 1
bracts soon deciduate
bud
fruit arrangements
leaf forms
1cm

A useful dense screening shrub for the garden.

dense, bushy shrub

M. densipicata Byrnes

DESCRIPTION ➤ Dense, woody shrub, normally 2–4m high by 2–2.5m wide, the bark grey, scaly or papery.
LEAVES small and narrow, glabrous, linear to narrowly ovate, about 3–13mm long by 1mm wide, acute, decussate and sessile.
FLOWERS white, in monads on many-flowered, dense spike to 40mm long; buds pink; hypanthium smooth and green; petals to 2mm long. Flowering season: spring–summer.
FRUITS form densely-packed, sessile spike about as long as inflorescence.

DISTRIBUTION ➤ Qld, on Western Darling Downs; common near Yelarbon in low-lying heavy soils and along main highway to near Goondiwindi. Also south to near Bourke in NSW.

DISTINGUISHING FEATURES ➤ Narrow, decussate, sharp-pointed leaves, tightly packed white flower spikes, and shrubby habit.

CULTIVATION ➤ Unknown to the author in cultivation but should be useful in difficult, low-lying, saline, heavy soils, either acidic or alkaline, where choice is limited. Dense growth could be utilised for windbreaks or hedging. Claimed to be popular in sub-tropical plantings.

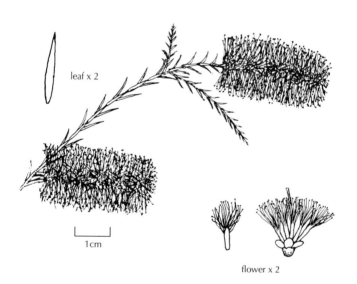

leaf x 2

1cm

flower x 2

A useful shrub for heavy, saline soils.

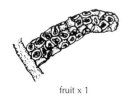

fruit x 1

dense, woody shrub

M. depauperata Turcz.

DESCRIPTION ➤ Dense, bushy shrub, 2m or more high and wide, with spreading branches.
LEAVES spirally arranged, usually 3–6mm long by 2–3mm wide, narrowly ovate or oblong, normally obtuse but sometimes pungent-tipped, fleshy but flat, spreading and glabrous.
FLOWERS mauve–pink to violet, fading to white with age, occurring profusely in leaf axils in short heads of 4–17 monads, each head on a short stalk. Flowering season: usually mid- to late spring.
FRUITS more or less spherical, in clusters; valves sunken; each capsule about 5mm long and wide.

DISTRIBUTION ➤ WA, from the Stirling Range and adjacent areas as far north as Wagin and east to Peake Charles, in forest, mallee or shrubland.

DISTINGUISHING FEATURES ➤ Narrowly ovate to oblong, spreading, widely spaced, small, fleshy leaves, and mauve–pink to violet flowerheads on short axillary stalks along the branches.

SIMILAR SPECIES ➤ *M. camptoclada* Quinn, localised to a small area in WA from Mt Barker to the Stirling and Porongorup Ranges, is almost identical but differs in its very recurved, greyish leaves with a more rounded apex and often strong margins.

CULTIVATION ➤ Both species are adaptable in areas of winter rainfall in excess of 400mm annually, but have proved difficult on the east coast from Sydney northwards.

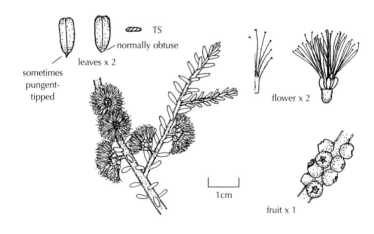

This shrub has superb flowers in good forms.

dense, spreading shrub

M. depressa Diels

DESCRIPTION ➤ Usually small, bushy or twiggy shrub, about 0.5m high and wide, but sometimes to 2m tall, particularly in semi-shaded locations.
LEAVES spirally arranged, about 6–12 mm long by 1.5–5.5mm wide, with 3–5 veins, elliptic to obovate or narrowly so, with sparse pubescent-like hairs, a tiny stalk and acute apex, and linear in transverse section.
FLOWERS in yellow or cream, capitate, terminal and upper axillary heads up to 20mm wide, comprising 2–6 triads; hypanthium hairy; petals deciduous; usually 9–13 stamens per bundle. Flowering season: mainly spring.
FRUITS mainly irregularly clustered (peg-like), mostly 3–4mm long, calyx lobes weathering away or forming sepaline teeth.

DISTRIBUTION ➤ Geraldton–Northampton district of WA, mostly in sand.

DISTINGUISHING FEATURES ➤ Elliptic to obovate, or narrow leaves of these shapes, flat, with sparsely pubescent hairs and acute apex, 6–12mm long; yellow to cream inflorescences to 20mm wide.

SIMILAR SPECIES ➤ *M. zonalis* Craven, from Eneabba to Gairdner Range of WA, where it is common on the laterite belt, is very similar, differing in its generally longer (up to 30mm), narrower, often spoon-shaped leaves and larger yellow flowerheads (to 28mm across). New growth is silky-grey and the fruiting clusters either globose or peg-like.

CULTIVATION ➤ Both species flower profusely, and could form attractive, small bushes for garden use with careful and regular pruning; they are probably suited to a range of soils in semi-dry to temperate areas. Both species are very successful in deep, well-drained soil.

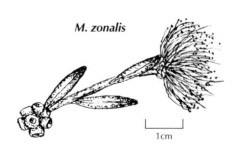
M. zonalis
1cm

These small shrubs flower profusely.
Best in well-drained soils.

M. zonalis

M. depressa

flower x 1

leaf x 1

1cm

small, bushy shrub

M. diosmatifolia Dum. Cours.

DESCRIPTION ➤ Normally a low shrub, 1–1.5m high, with spreading branches.

LEAVES sub-terete or linear, glabrescent, 4–15mm long by under 1mm wide, with small, non-prickly mucro; spirally arranged and crowded, in flat layers more or less parallel to branches; oil glands distinct.

FLOWERS pink or purple, in cylindrical spikes up to 40mm long; stamens 15–26 per bundle; new leaves growing on from end of spike. Flowering season: usually November–December.

FRUITS small, cup-shaped capsules, about 3mm diameter, clustered into a spike.

DISTRIBUTION ➤ Western slopes and ranges of NSW, and Darling Downs in Qld, favouring sandy depressions subject to flooding. Occurs in the Blue Mountains and adjacent and coastal areas of the Sydney region.

DISTINGUISHING FEATURES ➤ Small, terete leaves with distinct non-prickly hook, often arranged in layers more or less parallel to the branches.

CULTIVATION ➤ One of the smaller *melaleuca* shrubs, in cultivation for many years (as *M. erubescens*). It has proved adaptable to most soils and climates where rainfall exceeds about 500mm annually. The reasonably long summer flowering period has been a factor in its popularity.

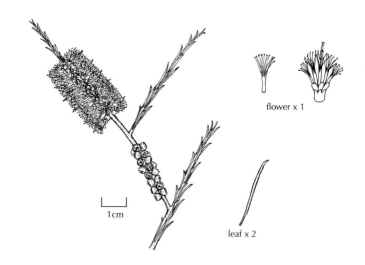

flower x 1

leaf x 2

1cm

A popular summer-flowering shrub.

low shrub

M. diosmifolia Andrews

DESCRIPTION ➤ Dense, imposing shrub, 1–4m high, with grey bark and unusual foliage, the new growth chartreuse in colour, at times giving a yellowish green appearance.

LEAVES alternate, stem-clasping, narrowly ovate, obtuse, reflexed and crowded thickly along branches; mature leaves about 10–12mm long by 5mm broad.

FLOWERS in cylindrical spikes 30–50mm long by about 40mm across, pale green to yellow–green, and tending to merge with foliage; stamens 3–5 per bundle. Flowering season: spring to early summer.

FRUITS woody capsules up to 10mm broad, generally clustered into a spike.

DISTRIBUTION ➤ Occurs along southern coastline of WA, from Albany to Cape Riche, growing amongst granite rocks, coastal heath and eucalyptus woodland. Naturalised in Otways district of Victoria.

DISTINGUISHING FEATURES ➤ Distinctive foliage and large greenish flower spikes.

The shrub known as the 'coastal form' of *M. diosmifolia*, with smaller, yellow flower spikes 20mm long by 18mm wide, is now recognised as a new species: *M. ringens* Barlow (see p. 246).

CULTIVATION ➤ Very adaptable shrub, widely grown in temperate Australia provided rainfall is above about 400mm annually. Can become ugly with age, so requires regular shaping to ensure an attractive bush prevails. Frost hardy.

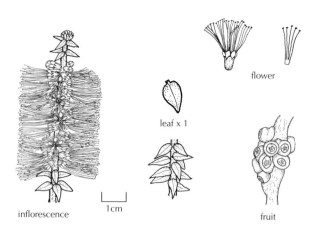

inflorescence | 1cm | leaf x 1 | flower | fruit

An imposing, dense shrub featuring greenish flowers.

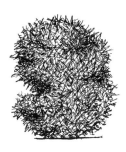

large, dense shrub

M. dissitiflora F. Muell.

DESCRIPTION ➤ Normally a tall, bushy shrub to 5m high, but may be erect or spreading, with grey, papery bark.

LEAVES spirally arranged, glabrous to glabrescent, except on pubescent young shoots, thin, narrowly oblanceolate to linear, normally 20–40mm long by 2–3mm broad, acuminate-acute and punctate-glandular.

FLOWERS creamy white, in loosely arranged spikes 20–60mm long, the axis growing on before flowers open; stamens pinnately arranged on claw, 15–35 per bundle; hypanthium hairy (often sparsely so); petals deciduous. Flowering season: usually winter and spring, but variable.

FRUITS sub-globular to flattened-spherical, about 3mm diameter, with infolded, persistent sepals.

DISTRIBUTION ➤ Inland species, found in Mount Chambers Gorge of the Flinders Ranges of SA, and elsewhere in north of SA, in NT and Qld, in sandy creek beds, rocky gorges and the like, with a disjunct occurrence in the Rawlinson Range of WA.

DISTINGUISHING FEATURES ➤ Thin, narrowly oblanceolate to linear, spirally arranged leaves, profuse cream flower spikes with pinnately arranged stamens, 15–35 per bundle.

CULTIVATION ➤ Rarely cultivated, but is very showy in flower and could be used as a large, ornamental shrub; particularly suited to low-rainfall gardens. Has been grown in the Adelaide foothills area where rainfall is about 700mm annually.

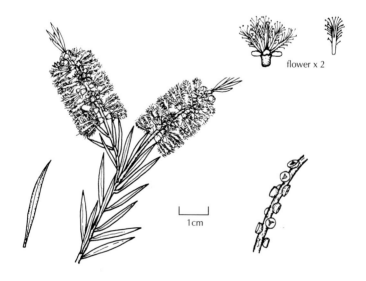

flower x 2

1cm

A showy, large shrub for low-rainfall gardens.

large, bushy shrub

M. eleuterostachya F. Muell.

DESCRIPTION ➤ Tall, glabrous shrub, generally 3–4m high, with thin, arching branches and greyish, papery bark.
LEAVES decussate, linear-lanceolate or linear elliptic, mainly 8–13mm long by 1–2mm wide, recurved, with hooked but not sharp mucro; variable species, with small-leaved forms not uncommon.
FLOWERS white or cream, in triads on short lateral spikes, about 20mm across by 15–25mm long; hypanthium glabrous; stamens 12–18 per bundle.
FRUITS spherical, 4–5mm diameter, with persistent sepals, in short, sessile, lateral spikes.

DISTRIBUTION ➤ Fairly widespread in central and northern Eyre Peninsula of SA, and inland WA, including the wheatbelt.

DISTINGUISHING FEATURES ➤ Short, lateral flower spikes and similar, characteristic fruiting spikes, and decussate, narrow, recurved leaves on thin, arching branches.

SIMILAR SPECIES ➤ *M. adnata* Turcz., occurring in WA from Kalbarri south and east to Ongerup and Mt Holland districts, in a wide range of soils and habitats, is similar. It differs in the narrowly ovate to narrowly elliptic, ovate, peltate leaves and also has some floral differences, including the base of the hypanthium being hairy.

CULTIVATION ➤ Both species are easy to grow in temperate areas of moderate to quite low rainfall, adapting to most soil types. They can be used for screening or a low windbreak but are not particularly ornamental.

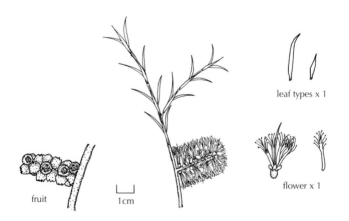

fruit 1cm leaf types x 1 flower x 1

M. eleuterostachya

M. adnata

These are tough shrubs suitable for most soils, including limestone.

M. adnata

flower

fruit

1cm

erect, woody shrubs

M. elliptica (Labill.) Kuntze
Granite Honey-myrtle

DESCRIPTION ➤ Medium-sized to large, woody, erect shrub, usually 2–4m high and less than 2m wide, with papery bark.
LEAVES decussate, blue–green to grey–green, elliptic to elliptic-ovate, 5–20mm long by 3–10 mm wide, on short stalks.
FLOWERS usually dark red; in large, showy lateral spikes to 80mm long by 50mm across; hypanthium hairy; other colour forms also described—yellow, white and pink, or light red. Flowering season: lengthy, mainly November–March.
FRUITS woody capsules about 8mm across, usually tightly compressed into a spike.

DISTRIBUTION ➤ Inland to coastal species from WA, found from Bendering and Ongerup districts to islands of the Archipelago of the Recherche, along the south coast extending almost to the edge of the Nullabor Plain. Favours granite outcrops and is sometimes known as Granite Bottlebrush.

DISTINGUISHING FEATURES ➤ Decussate, blue–green to greyish, elliptic to elliptic-ovate leaves, and large, cylindrical, usually dark red to pink, lateral flower spikes.

CULTIVATION ➤ Commonly cultivated. Can become woody and unattractive if not pruned each year once the required size is reached. Adapts to most soils and situations other than the tropics, except perhaps very wet, swampy soils, and is long flowering. Frost hardy.

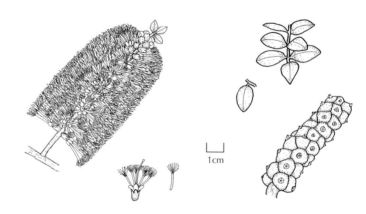

A popular large, long-flowering shrub.

large, woody shrub

M. ericifolia Smith
Swamp Paperbark

DESCRIPTION ➤ Tall, erect, bushy shrub, more usually a small tree, 6–9m, with a bushy crown and greyish, papery bark. Often suckers freely from the base.
LEAVES spirally arranged, glabrous, narrow-linear, 3–10mm long by 1mm or less in width, recurved and acute.
FLOWERS creamy white, in crowded, ovoid to cylindrical spikes 20–25mm long, the axis growing on; petals deciduous; stamens 7–14 per bundle. Flowering season: normally spring.
FRUITS cup-shaped, 3–5mm diameter, with raised valves.

DISTRIBUTION ➤ Common in low-lying swamps in Tasmania and King Island, where it grows in thickets. Once very common around Port Phillip Bay and elsewhere in Victoria, but few specimens now remain. Extends to NSW and Qld, mainly in wet coastal situations, including Sydney area.

DISTINGUISHING FEATURES ➤ Narrow, linear, recurved leaves, papery bark and crowded heads of whitish flowers. This species is variable and the leaves and fruits sometimes resemble those of *M. armillaris* (p. 24). It differs, however, in its papery bark and smaller, more prolific flowerheads.

SIMILAR SPECIES ➤ *M. parvistiminea* Byrnes, from the Shoalhaven River district in NSW south and west to Seymour district of Vic., and naturalised in south-west Victoria, is a shrub or small tree which also grows in thickets in wet places. It differs from *M. ericifolia* in its hard, rough bark, non-suckering habit, and longer, thinner flower spikes.

CULTIVATION ➤ Both species adapt well to cultivation wherever there is adequate water, and are excellent plants for swampy soils, and for hedging.

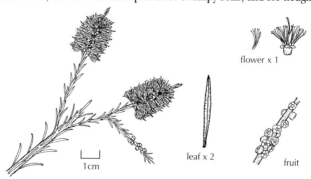

An adaptable tree suited to wet soils and good for coppicing.

slender, erect tree

M. eximia (K.J. Cowley) Craven

DESCRIPTION ➤ Erect, rigid, woody shrub, 2–2.5m high.
LEAVES subulate to linear-elliptic, to 20mm long, concave, peltate, decussate, crowded and sharply pointed; young growth softly hairy.
FLOWERS scarlet showy spikes 50–60mm long from leaf axils; large, very furry, cordate bracts subtend flowers; hypanthium densely hairy, lobes less so. Flowering season: long period from October into summer.
FRUITS in tightly clustered, long narrow spikes.

DISTRIBUTION ➤ Restricted to Mt Burdett–Wittenoom Hills area near Esperance, WA, in sandy soils.

DISTINGUISHING FEATURES ➤ Crowded, narrow, decussate leaves and showy, large, scarlet, axial flower spikes subtended by large furry bracts.

SIMILAR SPECIES ➤ *M. penicula* (K.J. Cowley) Craven, a rare species found only in Fitzgerald River National Park in southern WA, is very similar but differs in its narrowly ovate leaves and larger staminal claws.

CULTIVATION ➤ *M. eximia* grows well in sand, loam or clay in semi-dry to temperate areas where the pH is under 7. Probably not suited to saline or highly alkaline soils. Prune lightly to maintain a good shape.
  M. penicula is unknown to the author in cultivation.

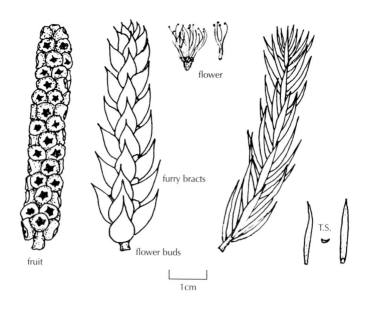

fruit / flower buds / furry bracts / flower / T.S.

1cm

An adaptable large shrub with impressive flowers.

M. penicula

slender, erect shrub

M. penicula

leaves x 1

M. fabri Craven

DESCRIPTION ➤ Small to medium-sized, open shrub, 1–2m high by 1m wide, with stiff branches.

LEAVES large, 60–90mm long by 10–15mm wide, leathery, glabrescent, narrowly elliptic to falcate, undulate with usually a sharp tip, and 3 prominent longitudinal veins; new growth silky-pubescent.

FLOWERS pink or mauvish pink, in large, cylindrical spikes up to 40mm long by 30–40mm wide, leaves growing on from tip of spike; rachis, hypanthium, lobes, bracts and petals all silky-hairy. Flowering season: spring (September–November).

FRUITS form an oblong conglomerate of variable size, capsules tightly packed together.

DISTRIBUTION ➤ WA, from Mullewa–Moora district to Wubin–Mt Gibson district, favouring sandy or gravelly laterite soils.

DISTINGUISHING FEATURES ➤ Large, leathery, striate leaves, coupled with large, pink to pinkish mauve, cylindrical flower spikes, with silky-hairy components.

CULTIVATION ➤ Unusual and very attractive shrub in full flower. Has proved easy to grow in clay or sandy, acid-neutral soils with reasonable drainage in areas of low humidity and dry summers. May not be suited to strongly alkaline soils or sub-tropical areas.

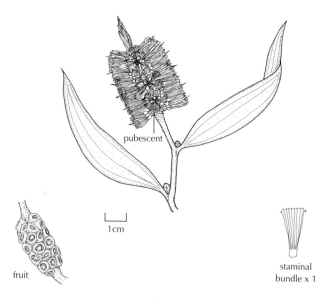

One of the most attractive flowering melaleucas.

stiff shrub

M. filifolia F. Muell.

DESCRIPTION ➤ Small, spreading or bushy shrub to 1m high and wide.
LEAVES needle-like, circular or nearly so in transverse section, mainly 15–40mm long, upcurved and spirally arranged.
FLOWERS bright pink, in globular heads 20–25mm across, terminal, but mainly on short side branchlets about 5mm in length, with consistently 5 calyx lobes; globular heads of greenish yellow buds always seem present in numbers matching flowerheads. Flowering season: October to summer.
FRUITS clustered into distinct 'soccer-ball' formation, 10–14mm diameter.

DISTRIBUTION ➤ Kalbarri–Mullewa district of WA.

DISTINGUISHING FEATURES ➤ Globular heads of bright pink flowers and greenish yellow buds, needle leaves to 40mm long, and 'soccer-ball' fruiting clusters.

SIMILAR SPECIES ➤ *M. boeophylla* Craven, also from Kalbarri district of WA, is very similar when in flower but differs in its much shorter leaves, 6–10mm long, semi-elliptic to semi-oblong in transverse section, and generally longer calyx lobes.

CULTIVATION ➤ Both species grow in acidic sand where annual rainfall is about 400mm but are relatively unknown in cultivation. *M. filifolia* is growing near Adelaide in well-drained sand where it thrives and should adapt to other soil types.

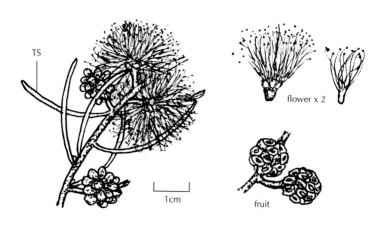

This small shrub has attractive flowers and buds.

M. boeophylla

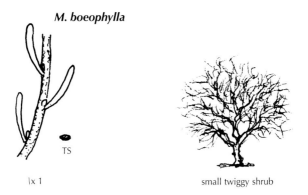

|x 1

small twiggy shrub

M. fulgens subsp. *corrugata*

(J.M.Black ex Eardley) K.J. Cowley

DESCRIPTION ➤ Small to large shrub, 0.7–3m, erect or spreading, with woody branches.

LEAVES decussate, narrowly elliptic with incurved margins, 10–25mm long by 1–4mm broad, acute; central vein and oil glands (on undersurface) prominent.

FLOWERS white to pale blush pink or mauve, arranged decussately in spike about 30mm long and wide on a lateral shoot; each flower with a thick stigma; stamens and claw shorter than in *M. fulgens* subsp. *fulgens*. Flowering season: Autumn to early spring.

FRUITS depressed-spherical to urn-shaped, 6–9mm diameter, in opposite pairs.

DISTRIBUTION ➤ Rocky hills of north-west SA and adjacent areas of WA and NT. Occurs in the Peterman, Musgrave and Rawlinson Ranges system, and in the Everard Ranges in SA.

DISTINGUISHING FEATURES ➤ Decussate, narrowly elliptic, pointed leaves with incurved margins and short, thick, lateral flower spikes, usually white or blush pink in colour.

CULTIVATION ➤ Not often seen in cultivation but has been grown successfully in the Adelaide Botanic Garden in good loam over limestone. Its rocky habitat suggests it would be best suited to free-draining soil in a dry to temperate climate.

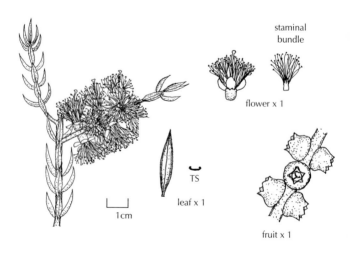

This is an excellent shrub for arid gardens.

woody shrub

M. fulgens subsp. *fulgens* (R. Br.) Cowley
Scarlet Honey-myrtle

DESCRIPTION ➤ Woody, usually erect shrub, 1–3m high by 1–2m wide.
LEAVES decussate, narrow, 10–40mm long, concave, with prominent oil glands, glabrous or glabrescent.
FLOWERS in large, lateral spikes 40–50mm across and long, colour varying from scarlet through pinkish red and pink to apricot; individual flowers decussate on spike; staminal bundles 20mm or more long, with numerous stamens arising from sides and front of claw. Flowering season: usually September–October, but spasmodic.
FRUITS flattened urn-shaped, 7–9mm across, in opposite pairs.

DISTRIBUTION ➤ Occurs from the Payne's Find district in WA south and eastwards to Israelite Bay and Great Victoria Desert.

DISTINGUISHING FEATURES ➤ Long, linear to linear-lanceolate, concave, leaves in opposite pairs, coupled with large, showy, scarlet, pink to apricot flower spikes.

CULTIVATION ➤ Easy to grow in semi-dry or temperate areas on most soils, but usually becomes chlorotic in limy soils. Popular in cultivation for its showy flowers over long periods. Can be kept quite small with regular pruning. Cuttings strike readily.

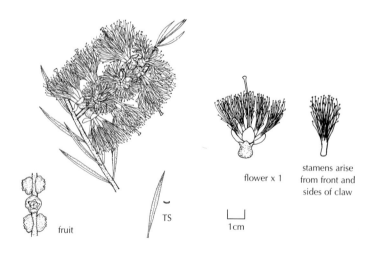

fruit

TS

flower x 1

stamens arise from front and sides of claw

1cm

This is one of the best melaleucas for cultivation.

erect shrub

M. fulgens subsp. *steedmanii*

(C. Gardner) Cowley.

DESCRIPTION ➤ Woody or twiggy shrub, usually to 1.5–3m, and normally erect, but may spread to 3m if unchecked.
LEAVES decussate, blue–green, glabrous, obovate to oblong or narrowly elliptic, up to 30mm long, with a small mucro.
FLOWERS occur laterally in showy crimson spikes 40–60mm long by 30–40mm across; flowers arranged decussately; golden anthers prominent; petals red. Flowering season: long period, mainly September–October but spasmodic.
FRUITS flattened urn-shaped capsules in opposite pairs.

DISTRIBUTION ➤ Northern sand plains of WA, from about Watheroo north to Wannoo.

DISTINGUISHING FEATURES ➤ Large, red flower spikes tipped with gold anthers, and blue–green, obovate to oblong, opposite leaves.

CULTIVATION ➤ Popular in cultivation because of its showy flowers over long periods and bluish green foliage. Control by regular pruning to keep it from growing too woody. Adapts to most soils in semi-dry to temperate areas, but resents lime.

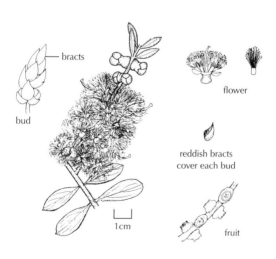

An adaptable, showy, long-flowering shrub.

woody shrub

M. fulgens R. Br., 'Orange-flowered form'

DESCRIPTION ➤ Woody shrub, usually smaller and less vigorous than the typical red-flowering form, normally 1m high and wide.
LEAVES decussate, narrow, mostly 10–30mm long by 1–2mm wide, concave, with prominent oil glands, and glabrous.
FLOWERS in showy, lateral spikes, apricot–orange to salmon-coloured; spike rounded and slightly smaller than in typical form; each claw supports numerous stamens. Flowering season: long period, mainly in spring.
FRUITS flattened urn-shaped capsules, in opposite pairs.

DISTRIBUTION ➤ Kalbarri National Park in WA, and possibly elsewhere.

DISTINGUISHING FEATURES ➤ Showy, apricot–orange to salmon-coloured, lateral flower spikes and generally smaller growth than the typical form.

CULTIVATION ➤ Popular in cultivation for its ornamental orange flowers over a long period. Slightly less adaptable than the other subspecies. Grows well in most acidic to slightly alkaline soils with good drainage in winter-wet climates of low humidity. Success outside this range is unknown to the author.

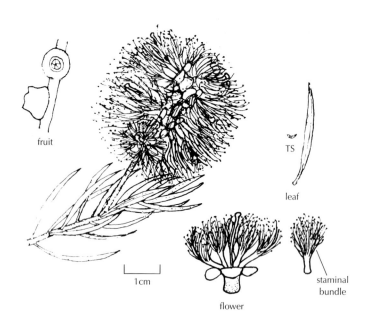

The unique flower colour of this melaleuca makes it a highly sought-after form.

woody shrub

M. 'Georgiana Molloy'

DESCRIPTION ➤ Erect shrub, 1.5–2.5m, with rough or scaly, greyish bark.
LEAVES needle-like, terete to sub-terete, fleshy, alternate, acute but not prickly, mostly 20–40mm long by 1–1.5mm wide, glabrous; oil glands prominent.
FLOWERS bright crimson with white style, in rounded heads 20mm across, crowded in clusters over long lengths of woody branches; young leaves protrude from flowerhead; hypanthium glabrous. Flowering season: November–December.
FRUITS form globular, woody heads along branches.

DISTRIBUTION ➤ Described as a naturally occurring hybrid from swampy bushland near Armadale, WA (1975), from which cuttings were taken by a WA nurseryman and introduced to cultivation as *M.* 'Georgiana Molloy'.

In the author's opinion and other's, this plant is a crimson-flowered form of *M. teretifolia* (p. 282).

DISTINGUISHING FEATURES ➤ Showy clusters of crimson flowers along branches, needle-like leaves and 'soccer-ball' fruiting clusters.

CULTIVATION ➤ Very attractive shrub in flower. Suited to temperate areas of assured rainfall and acidic to neutral, well-drained soils. Fails in very heavy soils and those of alkaline reaction.

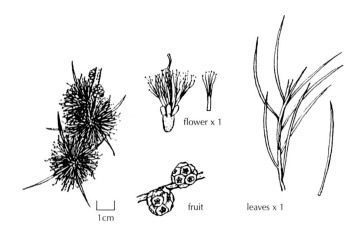

A spectacular shrub when it is in flower.

medium to large woody shrub

M. gibbosa Labill.

DESCRIPTION ➤ Attractive-foliaged, glabrous shrub, dense and bushy, with slender, arching branches, normally 1–2m high by about 2m across.
LEAVES sessile, decussate, ovate to obovate, concave above and keeled below, mostly 2–7mm long by 2–4mm wide, crowded along branches.
FLOWERS mauve to pink, fading to white as they age, in small, cylindrical or ovoid spikes or heads about 15mm long; flowers arranged in decussate pairs; spikes form at base of new lateral shoots, or along previous season's branches. Flowering season: mainly November–December.
FRUITS woody capsules about 5mm across, wider at base where they merge into thickened, woody rachis, a distinctive feature of this species and the similar *M. decussata* (p. 84).

DISTRIBUTION ➤ Common on Kangaroo Island, SA; also found in damp places in south-east of SA, in southern Vic. and Tasmania.

DISTINGUISHING FEATURES ➤ Distinctive fruiting arrangement, small, ovate to obovate, decussate leaves, and small mauve to pink flower spikes. Its normally lower growing, more spreading habit also separates it from *M. decussata* (p. 84). Intermediate forms with *M. decussata* are known, and several other forms, including one with very tiny leaves and a dwarf shrub with good groundcover.

CULTIVATION ➤ Adaptable, easily grown shrub, attractive in flower. Can be grown in most soils in temperate areas of moderate to quite high rainfall and is easily managed with regular pruning. Probably unsuited to very alkaline soils.

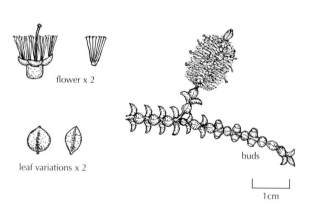

flower x 2

leaf variations x 2

buds

1cm

This is a good garden shrub for temperate regions.

fruit x 1

bushy shrub

M. glaberrima F. Muell.

DESCRIPTION ➤ Dense, glabrous, bushy, spreading shrub, 1–2.5m high and wide.

LEAVES spirally arranged and crowded, terete or flattened, often curving upwards, mostly 3–10mm long, acute or obtuse.

FLOWERS profuse, mauve to pink, soon fading to white, in short, lateral, cylindrical or oblong spikes about 15–20mm long, on old wood. Flowering season: normally November–December.

FRUITS small, cup-shaped, clustered into cylindrical spike.

DISTRIBUTION ➤ WA, from the Stirling Range to Cape Arid and inland to near Coolgardie.

DISTINGUISHING FEATURES ➤ Glabrous, needle-like but not prickly, terete, crowded leaves usually curving upwards. There are small and large-leaved forms.

CULTIVATION ➤ Easily grown shrub which does well in fairly heavy clay, loam or sand but resents high alkalinity. Suited to temperate or semi-dry areas where rainfall exceeds about 300mm.

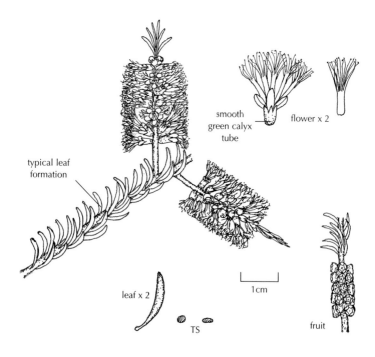

A free-flowering shrub for most temperate conditions.

bushy, spreading shrub

M. globifera R. Br., in Ait.

DESCRIPTION ➤ Normally a small tree, 6–10m, with light brown, papery bark and bushy crown.
LEAVES large, 30–70mm long by 10–20mm wide, smaller on flowering shoots, flat, thick and glabrous or glabresecent, obovate to oblong, usually slightly pointed with 5–7 or more parallel nerves.
FLOWERS cream to pale yellow, in globular terminal heads 30–35mm across; calyx lobes short, brown, often faintly ciliate; hypanthium glabrous. Flowering season: spring to early summer.
FRUITS form globular conglomerate of many capsules, normally about 20mm diameter.

DISTRIBUTION ➤ Eyre coastal region of WA, especially near Esperance, where it is particularly common at Lucky Bay–Cape Le Grande National Park. Also occurs on islands of the Archipelago of the Recherche.

DISTINGUISHING FEATURES ➤ Large, flat, obovate to oblong leaves with prominent longitudinal nerves; globular, cream to pale yellow, terminal flowerheads; globular clusters of fruiting capsules about 20mm in diameter.

SPECIES WITH SIMILAR FLOWERS ➤ *M. hnatiukii* Craven, found just south of Wittenoom Hills north to Scaddan area of WA, also near Esperance, favours low-lying depressions, or the edges of salt lakes, in sand. Although unrelated, this medium to large shrub has arching branches, narrowly elliptic, pungent-tipped leaves, and similar but smaller (about 20m wide) cream or white, globular flowerheads.

CULTIVATION ➤ *M. globifera* has been grown to a limited extent, adapting well to a range of non-limy soils, from heavy clay to sand, in temperate areas of assured rainfall. The author has no knowledge of its performance in limestone or highly alkaline soils. Habitat suggests it may be useful for foreground coastal planting.

M. hnatiukii is unknown in cultivation, but is obviously salt tolerant.

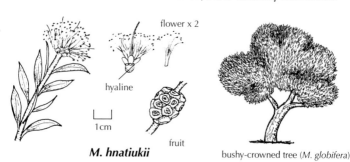

flower x 2
hyaline
1cm
fruit
M. hnatiukii
bushy-crowned tree (*M. globifera*)

M. hnatiukii

An adaptable tree well suited to coastal planting.

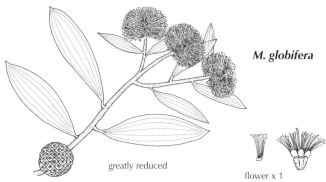

M. globifera

greatly reduced

flower x 1

M. glomerata F. Muell.
Desert Honey-myrtle, Inland Paperbark

DESCRIPTION ➤ Small tree or shrub, 3–10m, with spreading or straggly branches and whitish, papery bark.
LEAVES grey, linear to narrowly oblanceolate, flat, with straight, acute tip, glabrescent to densely appressed-hairy, 12–70mm long.
FLOWERS profuse, creamy yellow, in short, stalked, globular axillary and terminal heads; stamens 4–9 per bundle. Flowering season: usually late spring to early summer, but can be irregular in the wild.
FRUITS small, 2–2.5mm diameter, truncate, in clusters.

DISTRIBUTION ➤ Inland Australia, along dry streams. SA, including the Flinders Ranges, WA, NT and western NSW.

DISTINGUISHING FEATURES ➤ Silky-grey, linear, flat leaves with a straight, sharp tip; white, papery bark; small fruits; profuse globular, creamy yellow inflorescences.

CULTIVATION ➤ Very adaptable tree. Can be grown in most areas of Australia, from dry to temperate, in most soil types including mildly saline. Would probably succeed on the more humid east coast. Has been used as a street tree in Port Augusta, SA.

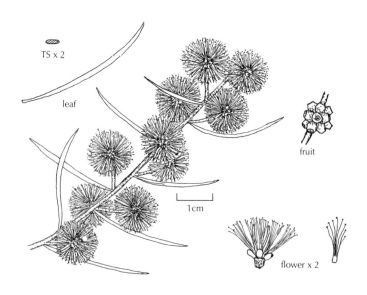

An adaptable and attractive-foliaged tree for most soils.

small straggling tree

M. groveana Cheel. & C.T. White

DESCRIPTION ➤ Large shrub to small tree, 5–10m high, straggly or bushy, with papery bark.
LEAVES spirally arranged, scattered and overlapping, narrowly elliptic to narrowly obovate, apiculate; 10–15mm long by 3–7mm wide on specimen illustrated but to 55mm long on some forms; glabrous with a short stalk; oil glands visible.
FLOWERS white; in open spikes 20–35mm long, arranged singly on spike; style longer than stamens (11–26 per bundle); hypanthium glabrous. Flowering season: mainly flowering September–October.
FRUITS spherical to cup-shaped, 4–5mm long and wide, singly spaced along branches.

DISTRIBUTION ➤ Uncommon, found in a few places in Moreton and Burnett districts of Qld (e.g. Mt Beerwah and near Chinchilla) and south to Port Stephens in NSW.

DISTINGUISHING FEATURES ➤ Smooth, crowded, elliptic to narrowly obovate leaves; quite large, white flower spikes, with flowers in monads.

ANOTHER UNCOMMON SPECIES ➤ *M. kunzeoides* Byrnes, from the Adavale district of dry central Qld, is a paper-barked shrub to 1.5m tall with somewhat similar leaves. It has yellowish green flower spikes, 4–6 stamens per bundle, and deciduating petals which are distinctly clawed.

CULTIVATION ➤ *M. groveana* is best suited to sub-tropical or warm temperate conditions; successful in a variety of soil types, flowering well in sun or shade. The author has no evidence of its tolerance to limy soils. It is also grown in the more southerly temperate areas such as Canberra. Frost hardy.
   M. kunzeoides is unknown to the author in cultivation.

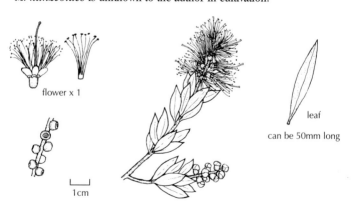

flower x 1

leaf
can be 50mm long

1cm

An attractive, white-flowered shrub.

M. halmaturorum F. Muell. ex. Miq.
South Australian Swamp Paperbark

DESCRIPTION ➤ Normally a tree of irregular or crooked growth, 4–8m, with thick, whitish, papery bark and dense crown; some trees resemble enlarged Japanese bonsai.
LEAVES small, dark green, glabrous (except on pubescent young shoots), decussate, linear-lanceolate, reflexed, with distinct glands on undersurface, 3–9mm long by 1–2mm broad, on a short stalk.
FLOWERS white or cream, in small axillary and terminal clusters, each cluster subtended by many concave, scarious bracts. Flowering season: spring.
FRUITS scaly, cup-shaped capsules, 3–4mm long and wide.

DISTRIBUTION ➤ Common in more temperate, low-lying parts of SA, particularly the south-east, in western Vic. and south-western WA. It favours saline, low-lying sites and is still found in these sites in the western areas of metropolitan Adelaide.

DISTINGUISHING FEATURES ➤ Normally crooked growth habit; thick, papery bark; small, blunt, decussate leaves and the numerous bracts subtending the inflorescence.

SIMILAR SPECIES ➤ *M. subularis* Barlow, from Peake Charles district of WA east to Zanthus, is a closely related small tree differing in its shorter leaves (2–4mm) and the few bracts subtending the inflorescence. Flowering season: September–October.

M. sparsiflora Turcz., from Marvel Loch district of WA south to the Oldfield River and east to Salmon Gums, is another related species, distinctive by its single-flowered inflorescence. Both it and *M. subularis* favour heavy soils on the edges of salt pans.

CULTIVATION ➤ *M. halmaturorum* is one of the very best plants for saline soils; in some very salty situations, it is the only species which is successful. It is used extensively to combat saline soils but will adapt to most other soils and conditions in temperate areas where rainfall exceeds about 400mm.

The other two species are not known to be in cultivation, but should also be useful for moderate conditions and saline soils.

This is an important species for saline soils.

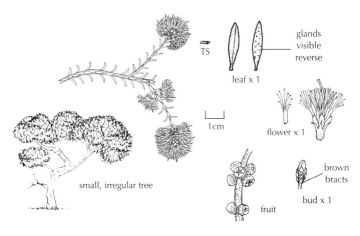

M. hamata Fielding & Gardner

DESCRIPTION ➤ Mostly a dense, large shrub 2–4m high by 2–3m wide, with papery bark which peels or flakes; occasionally a small tree to 5m.
LEAVES alternate, linear, needle-like, 20–80mm long by 1–1.5mm wide, glabrescent, ascending or spreading-ascending, straight or incurved, with a sharp, uncinate or straight tip.
FLOWERS profuse, cream to pale yellow, in globular heads along younger branches from almost every node; heads to 20mm wide; hypanthium usually smooth and green; stamens 3–8 per bundle; styles prominent, to 11.5mm long; the 5 distinct calyx lobes smooth on the outer surface. Flowering season: October–December.
FRUITS closely packed, usually into an ovoid cluster longer than its width.

DISTRIBUTION ➤ WA, from Mt Gibson–Lake Moore district southwards to Nyabing–Munglinup district and eastwards to Mt Ridley and the Wittenoom Hills.

DISTINGUISHING FEATURES ➤ Bulky but very showy, yellowish flowering shrub with more or less smooth, broombush-type, needle leaves 1–1.5mm wide, and distinct calyx lobes, smooth on the outer surface.

SIMILAR SPECIES ➤ *M. osullivanii* Craven & Lepschi, from Perth to Busselton districts of WA, is a similar, often smaller shrub which features thinner leaves, less than 1mm wide.

CULTIVATION ➤ Neither species often cultivated, although *M. hamata* is thriving in deep sand near Adelaide, where the accompanying pictures were taken. Both should grow well in well-drained soils in dry to moderate, temperate areas.

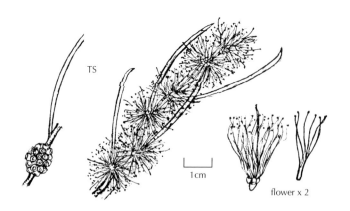

flower x 2

An outstanding species in full flower.

M. hamulosa Turcz.

DESCRIPTION ➤ Large, bushy, broom-like shrub 3–4m high and wide, with numerous ascending branches.

LEAVES spirally arranged, sub-terete with a pronounced hook, more or less adpressed, mainly 5–8mm long by 1mm broad, glabrous, oil glands prominent.

FLOWERS white or pinkish mauve, in spikes 20–50mm long by about 15mm across; hypanthium glabrous; axis growing on. Flowering season: usually late spring or early summer.

FRUITS flattened-spherical, compacted into a long spike along branches.

DISTRIBUTION ➤ Widespread in southern part of South-West Province of WA, mainly in areas with rainfall below 550mm annually.

DISTINGUISHING FEATURES ➤ Smooth, rather fleshy, adpressed leaves with a distinct hooked point; long flowering spikes with glabrous hypanthia. Appears to have hybridised with *M. armillaris* in cultivation.

CULTIVATION ➤ Very adaptable species, often used for screening because of its dense habit. Will grow in most soils, including mildly saline, as well as coping with dry to relatively winter-wet climates. Frost hardy.

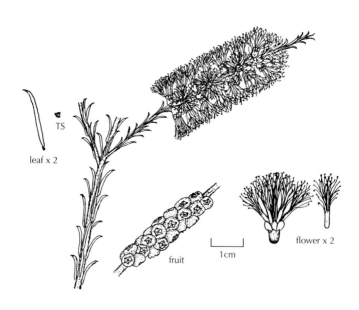

Excellent for windbreaks or screening.

Mauve form with lovely flowers.

dense, broom-like shrub

M. hollidayi Craven

DESCRIPTION ➤ Low shrub, mostly under 80cm, but spreading to 1m or more, with peeling bark.

LEAVES linear, to narrowly elliptic, 5–12mm long by 0.5–1mm wide, spirally arranged, warty, incurved or appressed, and mostly without stalks; branchlets and leaves feature cobwebby hairs.

FLOWERS profuse, rich pink with creamy yellow anthers, in heads to 22mm wide, of 2–9 triads; hypanthium woolly-hairy; stamens 5–10 per bundle. Flowering season: August–November.

FRUITS cup-shaped, 3–4mm long by about 3mm wide, in peg-like clusters, calyx lobes weathering away.

DISTRIBUTION ➤ Kalbarri–Northhampton–Mullewa–Mingenew districts of WA, in sand or loam over laterite.

DISTINGUISHING FEATURES ➤ Small, mostly, incurved leaves with cobwebby hairs; rich pink inflorescence with woolly hypanthium.

SIMILAR SPECIES ➤ *M. calyptroides* Craven, from WA, is of much wider distribution, from Watheroo–Moora district south and east to Merredin–Hyden–Coolgardie districts, on a variety of soil types. It is another mainly low, ground-hugging shrub. Leaves are linear, 6–28mm long by 0.7–1.7mm wide, distinctly warty and greyish, the younger ones silky-hairy. It is easily distinguished by its pink flowerheads which appear singly, or in 1–2 triads, hypanthium typically long (to 4mm), narrow and hairy, with scarious sepals. Flowering season: September–November.

CULTIVATION ➤ *M. hollidayi* is being grown successfully in Adelaide on deep sand and neutral clay-loam.

M. calyptroides is unknown to the author in cultivation but because of its wide range should be even more adaptable than *M. hollidayi*.

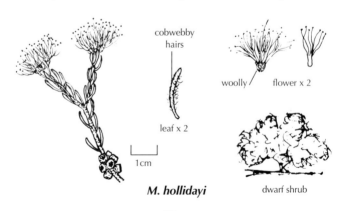

M. hollidayi — dwarf shrub

M. hollidayi

M. calyptroides

Both are free-flowering, spreading, low shrubs.

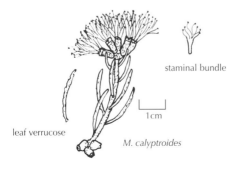

leaf verrucose

staminal bundle

1cm

M. calyptroides

M. huegelii Endl. subsp. *huegelii* Enum.
Chenille Honey-myrtle

DESCRIPTION ➤ Large shrub or small tree to 5m, with many intricate, straggly branches.

LEAVES peltate, ovate to almost triangular, acuminate, 2–5mm long by about half this in width, glabrous, sessile, and almost scale-like on smaller branches.

FLOWERS white, in pseudoterminal spikes to 100mm or more long in candle-like clusters; buds purplish pink; new leaf growth protrudes from ends of spikes. Flowering season: early summer.

FRUITS 3–4mm across, clustered along branches.

DISTRIBUTION ➤ Occurs on coastal limestone in WA, extending from Dirk Hartog Island to Augusta.

DISTINGUISHING FEATURES ➤ Numerous clusters of white, candle-like flower spikes in early summer; small, peltate, ovate leaves.

M. huegelii subsp. *pristicensis* Barlow is restricted to Dirk Hartog Island and southern end of Shark Bay. It particularly differs in its terminal and upper axillary, bright pink flowerheads, each about 20mm wide.

CULTIVATION ➤ *M. huegelii* is a large, often untidy shrub, ideal for coastal limestone soils and most others where drainage is good. It can be spectacular in flower, especially the form with purple buds which combine so well with the white flower spikes. Requires regular pruning to maintain its shape. Frost hardy. There is also a pink-flowering form in cultivation.

Subsp. *pristicensis* has rarely been cultivated.

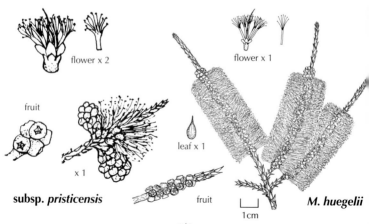

Subsp. *pristicencis* Pink flowering form

Profuse flowering shrubs for coast, limestone and other soils.

large shrub

M. buttensis Craven

DESCRIPTION ➤ Twiggy, brittle, often rounded shrub, 1–2m high and wide.

LEAVES elliptic, mainly 4–6mm long by 3–5mm wide, glabrous, or sometimes having a few hairs, fleshy, spirally arranged, obtuse or rounded at apex, and crowded, particularly towards ends of branchlets; longitudinal central nerve barely visible; oil dots faint; branchlets often bare within the shrub.

FLOWERS in terminal, globular heads; white with large yellow anthers, giving impression of being yellow; usually 8–12 stamens per bundle; hypanthia have short hairs; yellow, globular buds subtended by brown bracts. Flowering season: long period from September to late November.

FRUITS urn-shaped, about 5–6mm long and wide, usually in rounded clusters but occasionally in singles.

DISTRIBUTION ➤ Northern sand heaths of Hutt River district of WA.

DISTINGUISHING FEATURES ➤ Fleshy, elliptic, mainly smooth leaves, crowded towards ends of brittle branchlets; globular, yellow–white, terminal flowerheads with short-haired hypanthiums.

CULTIVATION ➤ Rarely seen in cultivation. Healthy specimens are growing at Koolunga in SA's mid-north, in light sandy loam. Could be useful in near-coastal plantings. With appropriate pruning makes an attractive small shrub with ornamental flowers.

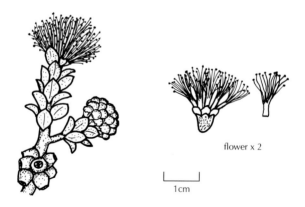

flower x 2

1cm

This melaleuca is worth trying in coastal gardens.

small, rounded, twiggy shrub

M. hypericifolia Smith
Hillock Bush

DESCRIPTION ➤ Large, normally erect, woody shrub or small tree, 3–6m, with handsome foliage and greyish, papery bark.

LEAVES glabrous, narrowly elliptic, more or less decussate (although a set of opposite leaves may not always be at right angles), more or less sessile; leaf size varies, from 30–40mm long by 6–14mm broad, to 10mm by 3mm on smaller branchlets.

FLOWERS orange–red to red, in large, cylindrical spikes 60mm long by 50mm in diameter, on short lateral stalks; singly or several together; partly hidden by foliage. Flowering season: usually late spring or early summer.

FRUITS tulip-like, woody capsules, 7–8mm in diameter when young, merging into a thick conglomerate of capsules as they age, with sepaline teeth.

DISTRIBUTION ➤ More temperate coastal and near-coastal regions of NSW, from Sydney area south to Bermagui.

DISTINGUISHING FEATURES ➤ Leaves large for a melaleuca, narrowly elliptic, smooth, in opposite pairs; large orange–red to red lateral flower spikes partly hidden by foliage.

CULTIVATION ➤ Has been cultivated for many years and is readily available from nurseries. Adapts to most soils and conditions, from subtropical to temperate. Useful as an informal hedge. A completely prostrate form also in cultivation makes an unusual groundcover.

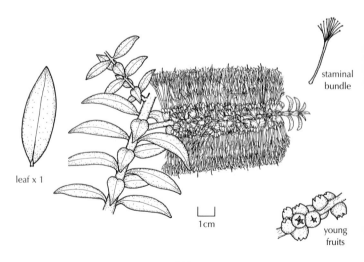

This popular paperbark has been cultivated over many years.

large, erect shrub

M. incana R. Br. subsp. *incana*
Grey Honey-myrtle

DESCRIPTION ➤ Medium-sized to tall shrub, 2–3m, with soft, weeping, hoary grey–blue to grey–green foliage and rough, dark grey bark.

LEAVES and smaller branchlets softly pubescent; leaves linear-lanceolate or linear-elliptic, 5–15mm long by 2–3mm broad, acute, curved, irregularly opposite or in whorls of 3, and crowded.

FLOWERS white to yellow, in soft, ovoid to oblong spikes, usually 20–30mm long by 20mm wide, freely produced. Flowering season: early spring.

FRUITS small woody capsules formed into tight, cylindrical spike normally about 30mm long.

DISTRIBUTION ➤ Wet, swampy places of WA's south-west, from near Albany to Jurien Bay area north of Perth. Common in damp situations at the foot of the Darling escarpment. Naturalised in parts of south-west Victoria.

DISTINGUISHING FEATURES ➤ Soft, hoary grey, weeping foliage.

CULTIVATION ➤ Commonly cultivated in temperate areas of assured rainfall. Adapts to most soils but resents high alkalinity. Frost tolerant. The dwarf cultivar 'Velvet Cushion' is popular amongst garden designers but is very prone to webbing caterpillar invasion in winter.

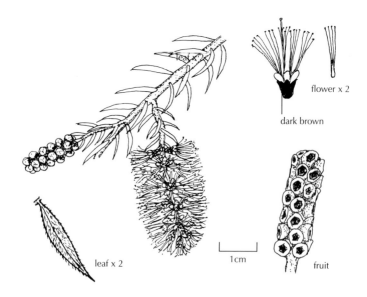

flower x 2

dark brown

leaf x 2

1cm

fruit

A popular shrub with outstanding foliage and habit.

weeping shrub

M. incana subsp. *tenella*

(Bentham) Barlow, in Quinn, Cowley, Barlow & Thiele

DESCRIPTION ➤ Erect to spreading shrub or small tree, to 5m, the slender branches arching and dainty.

LEAVES tiny, up to 9mm long, soft, dagger-shaped, flat, alternate or in whorls.

FLOWERS white, strongly scented, in oblong to cylindrical heads, 10–25mm long; hypanthia described as silky-pubescent by Blackall & Grieve (1980), but as seen by the author, glabrous and reddish. Flowering season: mainly October.

FRUITS thickly clustered into cylindrical spike to 20mm long, each capsule 3–4mm across and cupular.

DISTRIBUTION ➤ South coast of WA, from Esperance district to Cape Arid.

DISTINGUISHING FEATURES ➤ Small, soft, dagger-shaped leaves and dainty, pendulous growth habit.

CULTIVATION ➤ Attractive-foliaged plant which thrives in winter-wet light soils, and probably others as well. Successfully grown in sub-tropical areas as well as temperate. Frost tolerant.

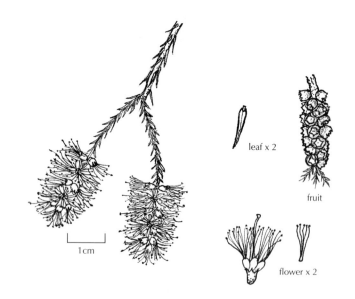

This plant is grown for its lovely soft, weeping foliage.

tall, graceful shrub

M. irbyana R.T. Baker

DESCRIPTION ➤ Small tree or shrub, 4–10m, with tiny-leaved, dense foliage and papery bark.

LEAVES 2–4mm long, elliptic to ovate or narrowly so, tapering to long narrow point, sessile, peltate and spirally arranged; branches display conspicuous cavities adjacent to each leaf blade.

FLOWERS white and scented, in few to many cylindrical spikes of 3–12 monads; hypanthia glabrous; stamens 6–11 per bundle.

FRUITS flat-spherical in clusters. Seed is shed annually.

DISTRIBUTION ➤ Moreton district of southern Qld, usually in pure stands or thickets, and favouring low-lying, poorly drained soils, south to Casino district and near Grafton in NSW.

DISTINGUISHING FEATURES ➤ Crowded, stem-clasping leaves, linear in transverse section, dense foliage and papery grey–white bark, with white flower spikes in monads.

SIMILAR SPECIES ➤ *M. tamariscina* Hook, from central-eastern Qld, has similar foliage, white flowers and papery, creamy brown bark. It is an ornamental, weeping tree with the flowering spikes comprised of mainly triads.

CULTIVATION ➤ Both species are well suited to sub-tropical areas, *M. tamariscina* requiring good drainage. Both have ornamental foliage and are grown as small tree specimens in Brisbane and surrounds. Their adaptability to temperate areas is not proven but specimens have succeeded in Adelaide.

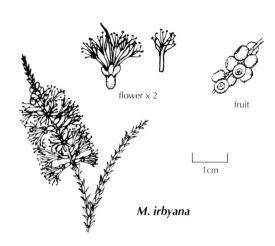

flower x 2

fruit

1cm

M. irbyana

M. tamariscina *M. irbyana*

Both species form attractive plants and are suited to the warm sub-tropics.

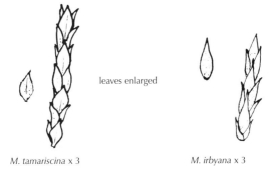

leaves enlarged

M. tamariscina x 3   *M. irbyana* x 3

M. laetifica Craven

DESCRIPTION ➤ Low, bushy shrub, usually under 1m high by about 1m wide, but may spread to 3m wide if unpruned.

LEAVES spirally arranged, linear to linear-obovate, normally 5–15mm long by 0.5–1.5mm wide, in transverse section circular to elliptic; warty with long, white hairs which are particularly thick on younger leaves.

FLOWERS bright yellow, in showy, rounded heads to 23mm across, held above foliage; heads comprise 4–10 monads; hypanthia silky-hairy; stamens 7–20 per bundle; petals deciduous, without oil glands; calyx lobes scarious throughout. Flowering season: October–November.

FRUITS urceolate to spherical, about 4mm wide, in peg-like clusters, calyx weathering away or replaced by sepaline teeth.

DISTRIBUTION ➤ Kalbarri–Hutt River district of WA, mostly in sand.

DISTINGUISHING FEATURES ➤ Narrow, hairy, warty, more or less terete leaves coupled with showy yellow flowerheads and peg-like fruiting clusters.

CULTIVATION ➤ One of the showiest yellow-flowering melaleucas; with its low, ground-hugging habit, can be used as a handsome foreground shrub or as a groundcover. It adapts well to acid-neutral sand, loam or clay where drainage is good, flowering best in full sun. Specimens occasionally die without warning. Suited to semi-dry to temperate areas but unknown in cultivation on the more humid east coast.

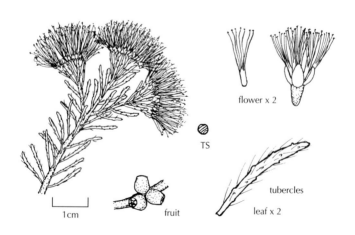

flower x 2

TS

fruit

tubercles

leaf x 2

1cm

Good groundcover with very showy yellow flowers.

low, bushy shrub

M. lanceolata (Otto) in Nees.
Moonah, Black Tea-tree

DESCRIPTION ➤ Dense, low-branching, dark-foliaged tree, usually 5–8m, with dark grey, rough bark; sometimes of shrub proportions in mallee woodland.

LEAVES crowded, 3–13mm long by 1–2mm wide, narrow, alternate, reflexed at apex, concave above, dark green (but new growth pale green).

FLOWERS white, in clustered cylindrical spikes 30–60mm long by 20mm across, flowers loosely spaced in spike. Flowering season: mainly summer, but can be irregular.

FRUITS smooth, woody capsules, spherical to ovoid, narrowed at the top, 5–6mm long and wide, loosely spaced along branches.

DISTRIBUTION ➤ From south-western WA eastwards to southern SA, western Vic. and NSW to south-central Qld.

DISTINGUISHING FEATURES ➤ Dense, dark green foliage and rough, dark bark, but other species have these features also and leaves may need close examination. Other distinctive features are light green new growth (usually in spring) and profuse, large, white flower spikes (usually in summer).

CULTIVATION ➤ Very adaptable tree, suited to difficult calcareous soils and coastal exposure, very attractive in full flower. Should be considered for street planting, especially for near-coastal streets and in areas of limy soil. May fail in the more humid east coast.

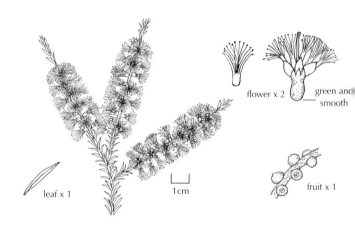

An excellent small tree for difficult soils and situations.

dense-crowned small tree of varying shapes

M. lasiandra F. Muell.

DESCRIPTION ➤ Handsome large shrub or small tree to 8m high, with papery bark, silky branchlets and silvery grey, attractive foliage.
LEAVES soft, silky-grey, flat, spirally arranged, crowded and overlapping, narrowly elliptic or narrowly obovate to falcate, mostly 10–50mm long by 1.5–10mm wide; tips acute to acuminate.
FLOWERS yellow to yellow–green or white; mostly in triads on spike 10–30mm long; hypanthia, rachis and staminal filaments softly hairy; stamens 6–20 per bundle. Flowering season: usually winter, but spasmodic.
FRUITS normally cup-shaped, about 2–3mm long by 3mm wide, in dense or open spikes.

DISTRIBUTION ➤ Tropical north, from northern WA including the Kimberley, through NT to just across border into central-western Qld.

DISTINGUISHING FEATURES ➤ Flat, narrowly elliptic to narrowly obovate or falcate, silky-grey, soft leaves and hairy staminal filaments.

CULTIVATION ➤ Used as an ornamental tree in Carnarvon, WA, and obviously suited to the drier monsoonal regions. The author has no knowledge of its cultivation elsewhere.

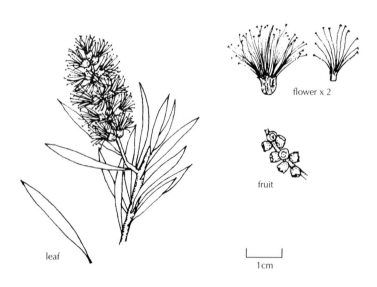

flower x 2

fruit

leaf

1cm

An attractive foliage plant for summer-rainfall areas.

large shrub or small tree

M. lateralis Turcz.

DESCRIPTION ➤ Two forms occur, the more usual a low, spreading shrub normally under 90cm, the other upright to 1.5m, with many narrow, ascending branches.
LEAVES tiny, mostly 2–4mm long, thick, somewhat fleshy, linear to narrowly elliptic or oblong, obtuse, faintly petiolate, spreading.
FLOWERS profuse, deep pink, in small, lateral, globular heads of monads on short stalks along branches; buds purplish; stamens 4–12 per bundle. Flowering season: early spring.
FRUITS small, smooth, cup-shaped capsules 3–4mm wide and long, with enclosed valves; lobes weathering away.

DISTRIBUTION ➤ Southern regions of WA's south-west, from Stirling Range district to Lake King district.

DISTINGUISHING FEATURES ➤ Tiny, obtuse leaves and clusters of pink flowerheads along the branches.

SPECIES WITH SIMILAR LEAVES & GEOGRAPHICAL OVERLAP ➤
M. lecanantha Barlow, in Barlow & Cowley, is an unrelated inland species, mainly from wheatbelt of WA in an area bordered by Wongan Hills, Southern Cross, Lake King and Nyabing. It is a shrub 0.2–2m tall, with similar leaves 4–7mm long, but ternately arranged, and pink to lilac flowers in the laterals comprising a single monad (sometimes 2) but many more stamens per bundle (19–30). Flowering season: September–October.

CULTIVATION ➤ Both species are adaptable dwarf to medium shrubs which grow well in most soils and situations in winter-rainfall areas of 350mm or more annually, and where drainage is good.

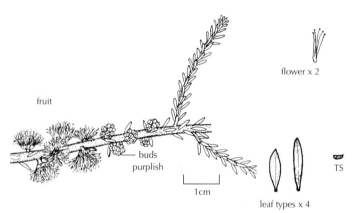

Attractive in flower and a good groundcover shrub.

low, spreading shrub

M. lateriflora Benth. subsp. *lateriflora*

DESCRIPTION ➤ Branching, normally erect shrub, 1–2.5m, but sometimes taller, to 6m.
LEAVES 4–12mm long by 3–6mm broad, flat, obovate, elliptic or oval on same bush, acute, and spirally arranged.
FLOWERS white, occurring laterally along branches in thick sessile clusters; hypanthia glabrous and green. Flowering season: normally early summer.
FRUITS 4–5mm diameter, with persistent sepals, in small, sessile clusters.

DISTRIBUTION ➤ Grows in a wide range of low to moderate rainfall areas in the South-West Province of WA, from near Geraldton to western edge of Nullarbor Plain.

DISTINGUISHING FEATURES ➤
Small, shiny, obovate to near oval leaves, mainly on upper branches, and lateral clusters of white flowers with smooth, green hypanthia.

M. lateriflora subsp. *acutifolia* (Benth.) Barlow ex Craven, occurring from Kalbarri south to Perth, features long leaves (to 25mm) which are not as flat in transverse section.

SIMILAR SPECIES ➤ *M. fissurata* Barlow, from the Hyden to Scaddan districts of WA, differs in its roughly textured, fissured, corky fruit, its keeled, retroflexed, shorter leaves and lower stature.
  M. eurystoma Barlow ex Craven, occurs in WA from the Corrigan district to Lake King and southwards to the Condingup district. It is related to *M. lateriflora* subsp. *lateriflora*, but differs by its smaller size, pale lemon to greenish yellow flowers, rounded leaf tips and many more ovules per locule (70–85 cf. 25–45).

CULTIVATION ➤ All are easily grown in semi-dry to moderate winter-rainfall temperate areas.

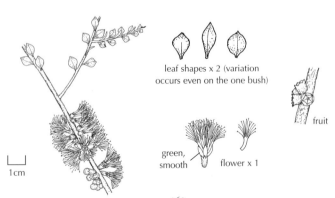

leaf shapes x 2 (variation occurs even on the one bush)

fruit

green, smooth   flower x 1

1cm

An adaptable large shrub for most temperate conditions.

erect shrub

M. lateritia A. Dietr.
Robin Redbreast Bush

DESCRIPTION ➤ Medium-sized shrub, normally 2–2.5m high by about 2–3m wide, with coarse, fibrous bark.

LEAVES spirally arranged, thin, linear, concave, glabrous except on the new shoots, 6–25mm long by 1–2mm broad, acute.

FLOWERS normally vivid orange–red, in large cylindrical spikes on lateral branches from old wood; spikes to 80mm long by 60mm across, with leafy shoots growing on from the axis. Flowering season: long period, peaking during summer and early autumn.

FRUITS rounded capsules 6–8mm diameter, with sunken valves.

DISTRIBUTION ➤ Widespread in WA, from Eneabba to Albany and many places in-between, including Perth region, favouring wet depressions.

DISTINGUISHING FEATURES ➤ Thin, linear leaves coupled with large, bright, normally orange–red flower spikes within the foliage.

SIMILAR SPECIES ➤ *M. apostiba* Cowley, a poorly known shrub from dry inland WA (Laverton–Lake Minigwal district), is closely related, but differs in its very hairy (tomentose) parts—leaves and floral features—as well as other characters such as its flat, shorter, narrowly elliptic leaves, 6–10mm long.

CULTIVATION ➤ *M. lateritia* is popular in cultivation for its vivid flowers and long flowering period. It favours areas of assured water and acid to neutral soils. Reliable as far north as Brisbane. Frost hardy.

M. apostiba is unknown to the author in cultivation.

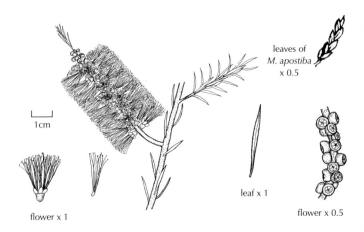

A popular garden shrub with spectacular flowers.

dense, medium-sized shrub

M. laxiflora Turcz.

DESCRIPTION ➤ Woody, usually spreading shrub, 0.5–3m high and wide, with rough, fibrous bark.

LEAVES glabrous, spirally arranged, narrowly elliptic to narrowly obovate, thick, 1–nerved, 10–30mm long by 1–4mm wide, with prominent oil glands.

FLOWERS vary from mauve, pink to white; sparsely arranged in cylindrical spikes of 6–20 monads; spikes 20–40mm long by about 20mm wide, numerous near base of lateral branches, the axis growing on; stamens 12–18 per bundle. Flowering season: normally late spring.

FRUITS cylindrical, 3–5mm long by 3mm across, wider at base, in loose clusters, with sepaline teeth.

DISTRIBUTION ➤ Inland WA, from Mollerin district south and east to Ongerup and Norseman districts, in varying habitats and soils.

DISTINGUISHING FEATURES ➤ Loosely arranged, normally lateral, mauve to pink flower spikes in monads; thick, oil-dotted, narrow, smooth leaves.

CULTIVATION ➤ Frequently cultivated shrub. Succeeds in most soils where drainage is good. Becomes chlorotic in soils where pH exceeds about 7.5. Suited to semi-dry and temperate climates and probably resentful of more humid areas.

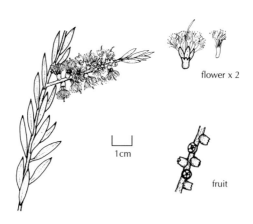

flower x 2

1cm

fruit

An attractive flowering shrub for semi-dry to temperate areas.

woody, spreading shrub

M. leiocarpa F. Muell.

DESCRIPTION ➤ Branching or erect, pungent-leaved shrub, 1–3m high.
LEAVES spirally arranged, 6–22mm long by 2–5mm wide, glabrescent except on pubescent young shoots, narrowly elliptic with pronounced acuminate-pungent apex, and with glands showing on both surfaces.
FLOWERS yellow or cream, in terminal or upper lateral spikes 15–25mm long and wide. Flowering season: usually spring to summer.
FRUITS woody, 7–9mm in diameter, and very spherical, resembling a miniature lawn bowl.

DISTRIBUTION ➤ Arid lands species, from Gawler Ranges and northern Eyre Pensinsula in SA and dry parts of eastern WA, including Coolgardie–Southern Cross area.

DISTINGUISHING FEATURES ➤ Narrowly elliptic to oblong, acuminate-pungent leaves and relatively large, very round fruits.

CULTIVATION ➤ Despite its prickly foliage, this is a very showy shrub in full flower, suited to most soils including mildly saline, in dry to temperate areas, but has rarely been cultivated. Should succeed in alkaline soils.

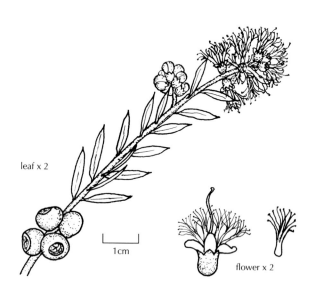

A showy shrub suited to dry sites.

erect or branching shrub

M. leiopyxis F. Muell. ex Benth.

DESCRIPTION ➤ Low, dense, slightly spreading or ground-hugging shrub, about 50 cm, or much taller and thinner, to 3m.

LEAVES slightly softly-hairy, spirally arranged, narrowly obovate to narrowly elliptic or linear-elliptic, the apex obtuse to pointed; slightly fleshy, 7–16mm long by 1–4mm wide, with a very short stalk; transverse section narrowly elliptic to linear; 3 parallel veins conspicuously thickened on underside; oil glands very obvious or sometimes obscure.

FLOWERS profuse, bright yellow and turning brown as they age; clustered in terminal and upper axillary, globular heads of 3–4 triads up to 25mm across; at their best providing a bright show over whole bush; hypanthia 2–3mm long and silky-hairy; stamens 8–14 per bundle; petals deciduous. Flowering season: mainly September–October but occasionally in winter, depending on the season.

FRUITS in small, irregular or peg-like clusters, each capsule about 5mm diameter by 4–7mm long, smooth and cup- or urn-shaped, the calyx lobes mainly weathering away.

DISTRIBUTION ➤ Confined to Kalbarri–Shark Bay areas of WA, growing in scrub, woodland, limestone ridges and beach dunes.

DISTINGUISHING FEATURES ➤ Leaves as described above and the bright yellow inflorescences comprised of 3–4 triads. This shrub is very difficult to distinguish from *M. depressa* (p. 92).

CULTIVATION ➤ Although rarely cultivated, this appears to be an excellent plant to trial in exposed coastal gardens, especially on alkaline dune systems and limestone soils. Should succeed in most soil types and could be a good shrub for difficult limy soils such as those encountered in many of Adelaide's northern districts. Grown successfully within Adelaide's West Lakes development.

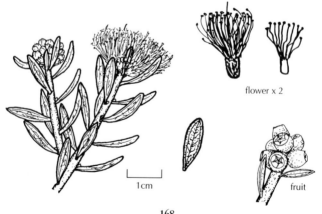

flower x 2

fruit

A very showy plant in flower worth a try in coastal gardens.

low, dense shrub

M. leucadendra (L.) L.
Weeping Paperbark

DESCRIPTION ➤ Large tree, normally under 20m but sometimes larger, with a thick, robust trunk supporting layers of thick, white, papery bark which is slippery to the touch. The smaller branches hang loosely downwards, giving a lovely weeping look, particularly to younger trees.
LEAVES thin, fresh green, lanceolate, narrowly ovate or falcate, to 250mm long by 6–40mm wide, with 5 or more distinct longitudinal veins or nerves.
FLOWERS cream or white, in loose cylindrical spikes 40–80mm long; fragrant, laden with nectar, and much sought after by birds and bees; stamens 5–12 per bundle. Flowering season: long, mainly winter to summer.
FRUITS cup-shaped, 4–5mm diameter, loosely clustered along branches.

DISTRIBUTION ➤ Widely distributed across tropical north, extending to Indonesia and New Guinea. A very common tree of the north Qld coast, growing right to the edge of the sea.

DISTINGUISHING FEATURES ➤ Attractive weeping habit of the thinner branches; thick, white, papery bark, and thin, fresh green, long leaves with 5 or more distinct nerves.

CULTIVATION ➤ Grown extensively as an ornamental and a street tree in tropical and sub-tropical areas. Common in Brisbane. There are a few healthy specimens in the Adelaide Botanic Gardens where abundant water is provided, indicating it can be a success in the colder south.

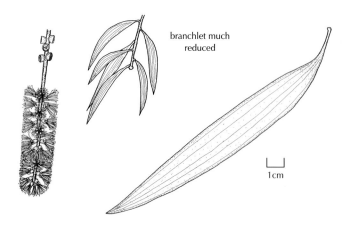

branchlet much reduced

1cm

A fine ornamental paperbark, especially those with a weeping form.

M. leuropoma Craven

DESCRIPTION ➤ Moderate-sized to small shrub, to 1.5m.

LEAVES glossy, with a few short, silky hairs which fall off rapidly once picked; spirally arranged, usually slightly warty, 4–15mm long by 1mm or so wide, with a small stalk or sessile; sub-circular to elliptic in transverse section.

FLOWERS purple or pink through to yellow, cream or white, usually brown in bud, mainly in terminal heads, the axis growing on; hypanthia silky-hairy; finger-like calyx lobes remain, varying from 2 to 4 or 5, or calyx a continuous ring of tissue; petals caducous, typically glossy. Flowering season: October–November.

FRUITS mainly cup-shaped, singly or in tight clusters, globose to irregular.

DISTRIBUTION ➤ From Kalbarri district to Gairdner Range–Moora district of WA, in sand.

DISTINGUISHING FEATURES ➤ Varying number of finger-like calyx lobes, which may vary from flower to flower on the same bush, and glossy petals, usually brown in bud.

CULTIVATION ➤ Unknown to the author in cultivation but well worth pursuing as it forms a most attractive, profusely flowering shrub. Should do well in well-drained acidic to slightly alkaline soils in semi-dry to temperate areas.

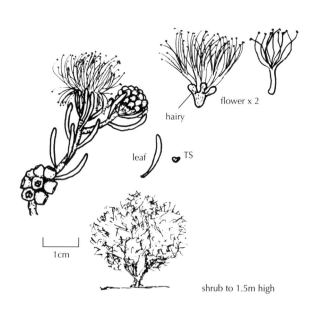

flower x 2
hairy
leaf
TS
1cm
shrub to 1.5m high

M. leuropoma (pink)   *M. leuropoma* (white)

A lovely ornamental species as yet untried in gardens.

M. linariifolia Smith
Narrow-leaved Paperbark

DESCRIPTION ➤ Low-branched tree, normally 6–10m, with a dense canopy of foliage, whitish, papery bark, and an abundance of flowers.
LEAVES soft, linear-lanceolate, 20–45mm long by 1–4mm broad, in opposite decussate pairs, with visible midrib.
FLOWERS snow-white to cream, prominently displayed in fluffy, massed heads over whole tree; stamens pinnate on the claw. Flowering season: late spring or summer.
FRUITS flattened-spherical, 4–5mm in diameter, with enclosed or sunken valves, calyx lobes weathering away.

DISTRIBUTION ➤ East coast, from about Ulladulla in NSW to Gladstone in Qld, with an isolated occurrence on Blackdown Tableland of Qld. Grows in swampy or dry woodland or open forest.

DISTINGUISHING FEATURES ➤ Massed heads of fluffy, white flowers dominating tree in late spring or summer, and fruits with sunken valves.

SIMILAR SPECIES ➤ *M. trichostachya* Lindl., in Mitchell, from southern NT, north-eastern SA eastwards into Qld and northern NSW, is a very similar small tree. It can be separated by its fruiting valves which are exserted or raised. The leaves may be spirally arranged or decussate.

CULTIVATION ➤ Both species will grow almost anywhere (except perhaps the northern tropics) where rainfall is in excess of about 400mm annually. However, strongly alkaline soils produce chlorotic leaves. *M. linariifolia* is used frequently in Melbourne as a street tree. It can be dramatic in flower but the flowering period is of short duration. Frost hardy. The cultivar 'Snowstorm' is a shrubby form, prolific in flower.

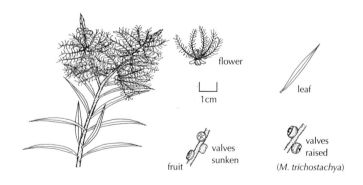

Spectacular in flower, this tree is very adaptable.

low, branching tree

M. linophylla F. Muell.

DESCRIPTION ➤ Straggly or bushy shrub, to 2–3m high, spreading to 2–2.5m across.

LEAVES narrow, glabrescent, spirally arranged, linear-oblanceolate, acuminate or narrowly elliptic, mostly 10–60mm long by 1–4mm wide, obscurely 1–nerved.

FLOWERS cream, in terminal and axillary spikes mostly 20–50mm long by about 18mm wide, the axis growing out before flowering; stamens pinnately arranged on claw. Flowering season: late spring (Adelaide), August–October in the wild.

FRUITS urceolate-globose, resembling a tiny vase, about 4mm diameter, sparsely arranged along branches.

DISTRIBUTION ➤ Karratha–Port Headland district of WA south to Paraburdoo district.

DISTINGUISHING FEATURES ➤ Narrow, flat leaves to 60mm long, numerous cream flower spikes and distinctly vase- or urn-shaped fruits. *M. dissitiflora* is very similar but differs in its more rounded, truncate fruits, as well as in less obvious floral features (see p. 98).

CULTIVATION ➤ Not well known in cultivation although a healthy specimen is growing at Wittunga Botanic Garden in the Adelaide Hills—an opposite climate to its natural summer-rainfall habitat. Should be a useful shrub for dry summer-rainfall regions.

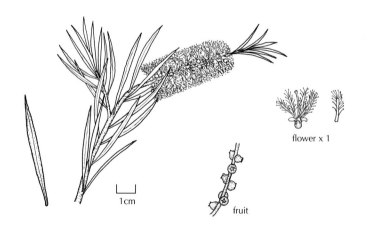

flower x 1

fruit

A useful shrub, especially for dry monsoonal areas.

2-m shrub

M. longistaminea subsp. *spectabilis*

Barlow ex Craven

DESCRIPTION ➤ Low, sprawling or prostrate shrub, under 1m high, with many spreading, woody branches.

LEAVES spirally arranged, glabrescent, stiff, ovate-cordate, cordate or ovate, peltate, pungent, with many fine striae, usually about 5–12mm long by 3–10mm wide.

FLOWERS lime green to yellow–green or yellow, clustered along branches, mostly in spaced, lateral heads to 45mm wide, of 5–15 monads; flowers subtended by leafy brown bracts and bracteoles 5–8mm long. Flowering season: normally September–late November.

FRUITS 5–6mm long and wide, with sepaline teeth, clustered along branches.

DISTRIBUTION ➤ Geraldine–Arjana district of WA.

DISTINGUISHING FEATURES ➤ Sprawling growth habit, pungent, heart-shaped leaves with many striae or veins, normally greenish yellow flowers subtended by brown bracts 5–8mm long.

M. longistaminea (F. Muell.) Barlow ex Craven subsp. *longistaminea*, from Murchison River–Carnamah district of WA, differs in its smaller flowerheads, much smaller bracts and bracteoles (mostly under 2mm long) and usually narrower, ovate leaves.

CULTIVATION ➤ Plants are not often seen in cultivation and are rarely available. Specimens have been grown successfully in Adelaide in acidic, well-drained sand over clay; performance in other conditions is unknown to the author.

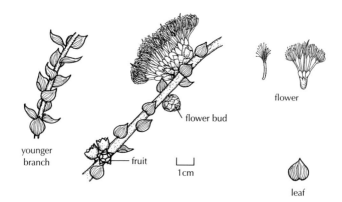

An unusual groundcover with interesting greenish flowers.

prostrate, woody shrub

M. macronychia Turcz. subsp. *macronychia*

DESCRIPTION ➤ Medium-sized shrub, normally 1–3m, and erect, with many branches.

LEAVES spirally arranged, glabrescent, elliptic, or lanceolate to oblanceolate, 10–30mm long by 2–15mm wide.

FLOWERS bright red, in large cylindrical spikes 40–60mm long by 30–50mm wide, on short lateral stalks; numerous stamens (15–25) clustered at end of staminal claw. Flowering season: begins about November and may continue to winter, depending on the season.

FRUITS spherical capsules about 6mm wide and long, clustered along branches.

DISTRIBUTION ➤ Inland WA, mainly amongst granite outcrops in Kalannie to Hyden districts.

DISTINGUISHING FEATURES ➤ Lanceolate to oblanceolate leaves, and large red flower spikes on lateral stalks.

M. macronychia subsp. ***trygonoides*** Cowley, occurring in WA further to the east, as far as Cave Hill, is distinguished by its shorter and more rounded, broadly elliptic leaves.

CULTIVATION ➤ Subsp. *macronychia* is a popular, long-flowering medium to large shrub with handsome flowers. Grows well in most acid to neutral soils with good drainage but struggles in strongly alkaline clays and limestone. Subsp. *trygonoides* should perform similarly but is unknown to the author in cultivation. Both are probably only suited to temperate areas.

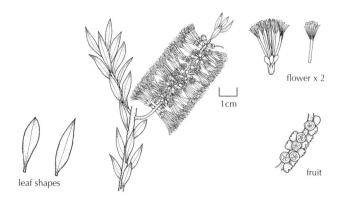

flower x 2

1cm

leaf shapes

fruit

This long-flowering shrub has large showy flower spikes.

erect shrub

subsp. *trygonoides*

M. manglesii Schauer, in Lehm.

DESCRIPTION ➤ Shrub, normally 60–70m high, spreading to about 1m wide.

LEAVES spirally arranged, recurved or spreading, glabrescent, narrowly obovate or narrowly elliptic, obtuse, to 10mm long but often less, by 1–1.5mm wide; visible oil glands on undersurface, more or less in rows.

FLOWERS profuse, deep pink, gold-tipped, in dense, capitate heads, singly or several together, each head 15–20mm wide; hypanthia hairy-white, hairs longer near base; brownish petals caducous; stamens 5–7 per bundle. Flowering season: spring.

FRUITS smooth, cup-shaped to urceolate, about 4mm wide and long in small, irregular clusters.

DISTRIBUTION ➤ WA, from Cowcowing Lakes district to Meckering–Kellerberin districts.

DISTINGUISHING FEATURES ➤ Almost smooth, mostly narrowly obovate, obtuse, spreading or recurved leaves with oil glands more or less in rows, and mostly under 10mm long. Fruits in small peg-like clusters.

SIMILAR SPECIES ➤ Two other WA species are very difficult to separate: *M. seriata* Lindl, from Green Head–Coorow district south to Perth and Bunbury, differs in its slightly longer petals which deciduate, often shorter leaves (to 4mm), sometimes weakly globose fruiting clusters and fewer ovules per locule.

M. parviceps Lindl., mainly from Darling Range–Perth area but also inland in Wyalkatchem–Kellerberrin district, differs by its longer (to 25mm), linear to linear-obovate leaves, generally wider inflorescences (to 25mm) and longer style.

CULTIVATION ➤ All three species (as *M. seriata*) have often been cultivated successfully in well-drained soils in temperate areas where rainfall ranges from 300–1000mm annually. Probably unsuited to the more humid east coast.

low, spreading shrubs

M. parviceps *M. seriata*

These groundcover shrubs are useful for a showy foreground.

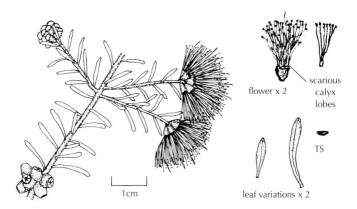

M. megacephala F. Muell.

DESCRIPTION ➤ Small to medium-sized, leafy, branching shrub, 1–3m, with stiff, brittle branches.

LEAVES spirally arranged, obovate to elliptic, usually under 25mm long by 4–10mm broad, deep green, with 3 prominent longitudinal veins, glabrescent except for softly hairy younger leaves and shoots.

FLOWERS profuse, in showy, globular heads to 40–50mm wide, appearing bright yellow due to the numerous yellow anthers, although stamens are white or cream; hairy, brown, leafy, imbricate bracts enclose flowers before opening; hypanthia hairy; stamens 12–16 per bundle. Flowering season: early to mid-spring.

FRUITS cup-shaped capsules, 3–5mm diameter, forming globular or peg-like clusters.

DISTRIBUTION ➤ Geraldton to Murchison River area of WA, mainly in sandy soils.

DISTINGUISHING FEATURES ➤ Large, globular, yellow flowerheads, conspicuous above the foliage, and broad, obovate to elliptic leaves.

CULTIVATION ➤ Ornamental garden shrub; has been in cultivation over a long period. Adapts well to most soils (not saline) in temperate areas but requires regular pruning after flowers have finished to ensure profuse flowering in following year. A smaller form in cultivation features all-yellow floral components and very shiny, smooth leaves. This form is one of the very best yellow-flowering melaleucas for most gardens in temperate areas.

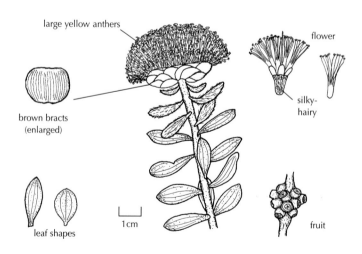

A spectacular yellow-flowering shrub for temperate gardens.

leafy shrub

M. micromera Schauer, in Lehm.

DESCRIPTION ➤ Tall, erect or spreading shrub, 1–3m, with many slender branches and short branchlets.

LEAVES minute, mainly in whorls of 3, broadly ovate, scale-like, fleshy, and peltately attached, foliage cypress-like in appearance; small branchlets usually covered with a close white tomentum which is concealed by the leaves.

FLOWERS profuse, yellow, in small, globular heads about 10mm wide, of 3–18 monads, but usually under 10. Flowering season: August–October.

FRUITS cup-shaped, about 3mm long and wide, usually in small clusters.

DISTRIBUTION ➤ Rare, found in WA in a small area near Mount Barker and in Tunney–Stirling Range–Green Range district, usually amongst jarrah.

DISTINGUISHING FEATURES ➤ Cypress-like foliage comprising broadly ovate, fleshy, minute leaves, about 1mm wide and long, in whorls of 3, and small yellow flowerheads.

CULTIVATION ➤ Unusual melaleuca, vulnerable in the wild and should be grown more to preserve the species. Has proved fairly reliable on non-limy soils in temperate areas but is sometimes prone to dying without warning.

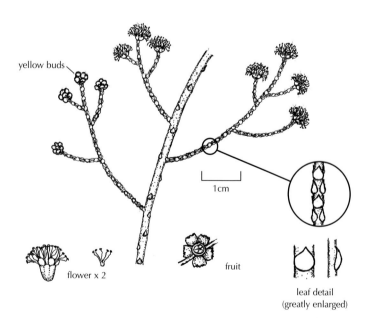

leaf detail (greatly enlarged)

A rare and unusual shrub that should be grown more often.

unusual shrub

M. microphylla Smith, in Rees.

DESCRIPTION ➤ Very dense, bushy, glabrous shrub, 2–4m, with handsome, narrow, bright green foliage and papery bark.
LEAVES glabrous, linear, normally 3–10mm long by 0.5–1mm broad, spreading and recurved, obtuse to slightly pointed; small-leaved forms are in cultivation.
FLOWERS white or cream, in profuse, rather narrow, cylindrical spikes 20–50mm long to 22mm wide; smell is sickly. Flowering season: usually spring.
FRUITS cup-shaped, 3–4mm diameter, clustered along branches.

DISTRIBUTION ➤ WA, from near Manjimup to Manypeaks district; common around Albany, particularly near coast.

DISTINGUISHING FEATURES ➤ Crowded, short, narrow, smooth, bright green leaves and dense growth habit. There appear to be several forms of this species, some with much larger flowers than those illustrated.

CULTIVATION ➤ Extensively cultivated for dense screening, hedging and coastal situations, where it will suffer windburn on the windward side. Succeeds in most soils in temperate areas where rainfall is above about 400mm annually, mainly in winter.

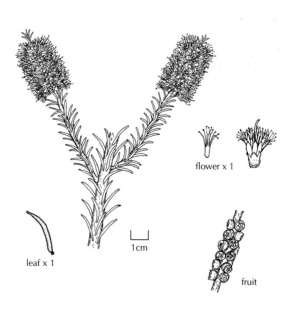

leaf x 1

1cm

flower x 1

fruit

An excellent shrub for hedging or screening.

dense, bushy shrub

M. sp. aff. *microphylla*

DESCRIPTION ➤ Slender, weeping, graceful tree to 6m high with white, papery bark.
LEAVES small and soft, glabrous, 3–8mm long, normally under 1mm broad, terete, recurved and obtuse, spirally arranged, and crowded on small branches.
FLOWERS bright yellow to yellow–green, in cylindrical spikes to 25mm long; often in thick clusters. Flowering season: usually October–November.
FRUITS more or less cup-shaped, clustered tightly into small cylindrical spike.

DISTRIBUTION ➤ Denmark–Walpole district of south-west WA, in wet places.

DISTINGUISHING FEATURES ➤ Graceful, drooping habit and small, soft leaves on slender branches. Although the leaves resemble those of *M. microphylla*, the weeping, slender habit, white, papery bark and yellow to yellow–green flower spikes separate it from the typical form of that species.

CULTIVATION ➤ Growing well at Wittunga Botanic Garden in the Adelaide Hills, but no evidence elsewhere. Has excellent landscape potential because of its relatively small size and weeping habit. Probably requires locations where water is assured.

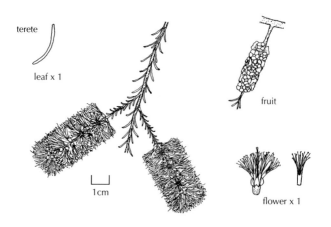

A lovely weeping paperbark tree for landscaping.

slender, weeping tree

M. minutifolia F. Muell.

DESCRIPTION ➤ Small tree or shrub, 2–10m, with many branches and white, papery or fibrous bark.

LEAVES minute, about 2mm long, rhombic or narrowly rhombic, acuminate, decussate, appressed, peltate with dark-coloured glands in raised pustules.

FLOWERS white or cream, in short spike in dyads; stamens 4–8mm long; surrounding bracts broadly ovate. Flowering season: flowers may occur at any time but more likely in autumn or spring.

FRUITS broadly cylindrical, 2–3mm long by 3–4mm wide, with persistent calyx lobes.

DISTRIBUTION ➤ Western Kimberley in WA to south-west Arnhem Land of NT.

DISTINGUISHING FEATURES ➤ Tiny, rhombic, decussate, pointed, pusticulate, peltate leaves, and flowers in dyads.

SIMILAR SPECIES ➤ *M. foliolosa* A. Cunn. ex. Benth., from central and southern Cape York Peninsula, is of a similar size range, the tree featuring a bushy canopy. Flowers are in monads. It differs from *M. monantha* by its acute leaf apex, appressed leaves (cf. ascending) and 20–30 stamens per bundle.

M. monantha, from the Palmer River to Mt Sturgeon in Qld, differs in its flowers, which are in monads, as well as its non-pustular leaf glands. The leaf blade is acuminate and there are 6–14 stamens per bundle.

M. nanophylla Carrick, from arid north-west SA, is similar but distinguished by its broadly ovate, mainly obtuse, non-pustular, smooth leaves, spirally arranged.

CULTIVATION ➤ Cultivation of these species is unknown by the author but both *M. monantha* and *M. foliolosa* could be used as unusual specimens for the wet tropics. The other two species may succeed in drier conditions.

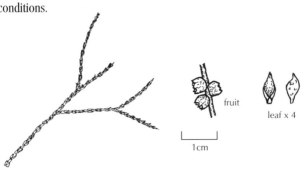

fruit

leaf x 4

1cm

M. foliolosa

M. minutifolia

Tiny-leaved shrubs or trees worthy of garden trial.

M. minutifolia

leaf arrangement greatly enlarged

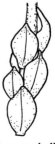

M. nanophylla

M. nematophylla F. Muell. ex Craven
Wiry Honey-myrtle

DESCRIPTION ➤ Normally an erect, rounded shrub, 1–3m high, with needle-like leaves and brownish white, papery bark.

LEAVES terete, spirally arranged, glabrous, 40–160mm long, curving upwards, apex sharply pointed but not prickly.

FLOWERS produced freely, in globular or ovoid terminal heads 50mm or more wide over whole bush; deep or pale pink and very showy; buds also attractive, appearing as reddish green, diamond-like heads due to colour of bracts covering unopened flowers. Flowering season: normally September–October.

FRUITS cup-shaped, 6–8mm diameter, usually formed into an irregular, peg-like cluster.

DISTRIBUTION ➤ Northern sand heaths and gravelly soils of WA, from Kalbarri to Three Springs–Perenjori district, and also Manning Range–Bungalbin district.

DISTINGUISHING FEATURES ➤ Long, needle-like leaves, showy, terminal, globular, deep to pale pink flowerheads and conspicuous buds.

CULTIVATION One of the most popular species in cultivation because of its showy flowers and ease of growth, although it becomes chorotic in very alkaline soils. Should be pruned back beyond the same year's flowers each year to maintain an attractive bush and ensure a continuity of showy flowering. Best suited to dry temperate to temperate areas.

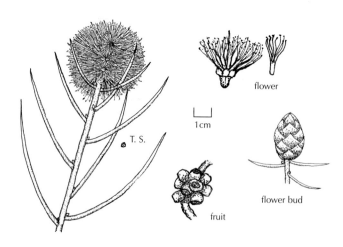

In flower, this is one of the most spectacular melaleucas. It adapts well to most temperate garden conditions.

erect, rounded shrub

M. sp. aff. nematophylla

DESCRIPTION ➤ Small to medium-sized, loosely branched shrub, usually 1–2m high with equal spread.
LEAVES narrow, concave, spreading, to 60mm long by 3–4mm wide, linear, with a raised midrib, acute, glabrous except for the pubescent new growth.
FLOWERS purple–pink, large, globular flowerheads, mainly 20–60mm diameter, the axis growing on; hypanthia pubescent over bottom half. Flowering season: October–December.
FRUITS tightly packed into globular or ovoid clusters.

DISTRIBUTION ➤ Unknown.

DISTINGUISHING FEATURES ➤ Narrow, concave, spreading leaves to 60mm long, and large, globular purple–pink flowerheads followed by tightly packed fruiting clusters.
  This plant is possibly a naturally occurring hybrid between *M. nematophylla* and *M. barlowii* (p. 64).

CULTIVATION ➤ Often cultivated in temperate areas. Grows well in acidic to slightly alkaline, well-drained soils, forming a smaller bush with similar flowers to the popular *M. nematophylla*.

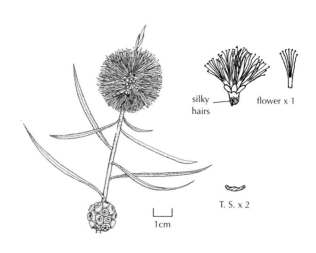

silky hairs
flower x 1
T. S. x 2
1cm

A showy, smallish shrub for temperate areas.

small, loose shrub

M. nervosa (Lindl.) Cheel. subsp. *nervosa*

DESCRIPTION ➤ Shrub or tree, 2–15m high, with erect branches, often untidy crown and layered, papery or fibrous bark, grey, cream, brown or white, or combinations of these.
LEAVES hairy and leathery, becoming smoothish (glabrescent), spirally arranged, elliptic to obovate, or narrowly or broadly so, apex acuminate and parallel veins (3–7) prominent; leaf size determining subspecies (see below); young growth very curly-hairy, linear in transverse section.
FLOWERS various shades of green to yellow, white or red; in large showy spikes of triads, spikes to 100mm long by 50mm wide; hypanthia and outer surface of calyx lobes densely hairy; petals deciduous; stamens 3–7 per bundle. Flowering season: can be erratic, usually winter to summer.
FRUITS cup-shaped, 2–4mm long and wide, scattered along branches in loose spikes; seed is soon shed.

DISTRIBUTION ➤ Widespread in Qld from about north of Bundaberg, through NT to northern Kimberley in WA, in varying habitats and often in pure stands.

DISTINGUISHING FEATURES ➤ Usually a small tree with papery bark, large, leathery, prominently veined, elliptic to obovate leaves with curly hairs overlaid by silky pubescent hairs which become sparse, slowly or rapidly.

M. nervosa subsp. *crosslandiana* (W. Fitz.) Barlow ex Craven, from the Kimberley in WA across central NT to north-west Qld, features leaves 5–30mm wide and 3–10 times as long, with hairs which soon disappear.

SIMILAR SPECIES ➤ *M. triumphalis* Craven, known only from Victoria River Gorge and associated gorges of NT, is a shrub to 2.5m. It is related but differs by its sub-papery, fissured bark, much longer petals (5–7mm cf. 1.5–3.5mm), and long, shaggy, branchlet hairs. Flowers are greenish.

CULTIVATION ➤ *M. nervosa* is grown successfully in many tropical and sub-tropical parts of Australia.
  M. triumphalis is unknown to the author in cultivation.

small tree

An attractive paperbark for warmer regions.

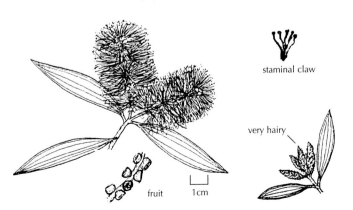

staminal claw

very hairy

fruit 1cm

M. nesophila F. Muell.
Showy Honey-myrtle

DESCRIPTION ➤ Large, dense shrub or small tree to 4m, with branches at or near ground level, normally erect and rather stiff, with greyish white, papery bark.

LEAVES smooth, elliptic, obovate or oblong-obovate, normally 10–20mm long but sometimes larger, bright green and spirally arranged.

FLOWERS conspicuous, deep violet to purplish pink, tipped with yellow anthers, sometimes massed thickly; in globular, mainly terminal heads 25–30mm across, singly or several together; hypanthia glabrous; stamens 7–14 per bundle. Flowering season: late spring and summer, occurring over long period.

FRUITS clustered into ovoid or round heads, 20mm or more long, but variable.

DISTRIBUTION ➤ South coast of WA (Eyre district, including Doubtful Island).

DISTINGUISHING FEATURES ➤ Bulky, dense growth habit, smooth, usually elliptic or obovate leaves, and conspicuous, purplish or pink flowerheads, with transparent margins to petals and calyx lobes.

SIMILAR SPECIES ➤ *M. glena* Craven, from Fitzgerald River to near Mt Burdett and Wittenoon Hills north-east of Esperance, WA, is a medium-sized shrub to 2–3m, with similar features to *M. nesophila* except for the purple inflorescences, which occur mainly in the leaf axils, rather than mostly terminally, are smaller, with hairy hypanthia and less stamens per bundle (4–8).

CULTIVATION ➤ *M. nesophila* is one of the most commonly cultivated melaleucas. It will grow almost anywhere, including harsh coastal sites in sand dunes, or on limestone. It is useful as a tough hedge or windbreak, recovering well from regular trimming. Adapts to the east coast. Frost hardy. The dwarf form known as 'Little Nessie' grows only to a bushy 1.5m.

M. glena is unknown to the author in cultivation.

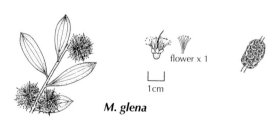

M. glena

flower x 1

1cm

M. nesophila

This is a popular shrub for coastal and most other situations and soils.

M. glena

large, dense shrub

M. nesophila

hyaline margins

flower x 1

1cm

fruit

leaf shapes

M. nodosa (Gaertn.) Smith

DESCRIPTION ➤ Shrub or tree, ranging from under 1m to 10m, with thin, arching branches.
LEAVES variable, from linear to near-terete, 10–30mm long by 0.5–15mm wide, spirally arranged with a sharp pungent tip.
FLOWERS profuse, deep yellow to white, in dense globular heads, 10–14mm across, terminally and in the leaf axils; hypanthia hairy; petals deciduous. Flowering season: any time from April to mid-summer.
FRUITS tiny, 2–3mm in diameter, cupular, sometimes a few well separated on the branch, but mainly in globular clusters.

DISTRIBUTION ➤ Fairly common on east coast, from southern Qld to Sydney area, in a range of soil types and habitats. Extends inland to tablelands.

DISTINGUISHING FEATURES ➤ Pungent, needle-like, but variable leaves 30mm long, tiny, cupular fruits, mainly in globular clusters, and profuse, yellow to white, globular flowerheads with hairy hypanthia.

SIMILAR SPECIES ➤ *M. borealis* Craven, from Lakeland Downs district of Qld south to Valley of Lagoons Station, is very closely related, differing in its longer leaves (about 20–50mm) and smooth hypanthia.

CULTIVATION ➤ *M. nodosa* is very adaptable both in soil requirements and location and is successful in most parts of Australia where water is assured. Tends to become straggly, requiring regular pruning. Showy in flower and frost hardy.
  The author has no knowledge of *M. borealis* in cultivation.

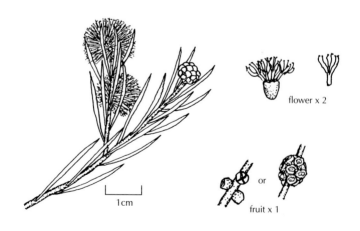

flower x 2

fruit x 1

This adaptable shrub or tree is showy in flower.

tall shrub or small tree

M. oldfieldii F. Muell. ex Benth.

DESCRIPTION ➤ Normally a loosely branched, leafy shrub, 1–2m high by about 1.5m wide, but reaches small tree proportions along streams where it grows naturally.

LEAVES spirally arranged, petiolate, light green, smooth and shiny, elliptic or obovate with a distinct mucro, mainly 15–20mm long by 4–8mm broad, with prominent longitudinal nerves; new shoots slightly hairy.

FLOWERS bright daffodil yellow, in large, globular terminal heads up to 35mm wide; buds also yellow, partly pubescent. Flowering season: normally November.

FRUITS in thick, globular clusters up to 25mm across, calyx lobes weathering away.

DISTRIBUTION ➤ Known only from Kalbarri National Park in WA, usually along stream banks.

DISTINGUISHING FEATURES ➤ Light green, shiny, pointed, elliptic to obovate leaves and intense yellow, globular flowerheads.

CULTIVATION ➤ Very handsome foliage and flowers, but tricky to grow. Success has occurred in acid to neutral soils with good drainage and where water is assured. A temperate climate plant which resents limy soils.

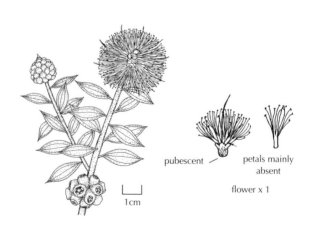

This shrub features beautiful flowerheads.

leafy, loose shrub

M. oxyphylla Carrick

DESCRIPTION ➤ Medium-sized, bushy shrub, normally 1–2m high and wide.

LEAVES evenly decussate, overlapping in a distinctive arrangement; individual leaves mostly 8–10mm long by 1–2mm wide, glabrescent, except for hairy new shoots, linear-lanceolate or elliptic, acuminate to acute, but not prickly, often recurved, with tiny stalk and faint oil glands.

FLOWERS white or cream, in lateral clusters of 1–5 monads along branches; hypanthia smooth; stamens 9–15 per bundle. Flowering season: spring.

FRUITS sub-globular or cup-shaped, smooth, 3–5mm in diameter, clustered mainly in groups along branches, with sepaline teeth.

DISTRIBUTION ➤ Endemic to Eyre Peninsula of SA, mainly in central wheatbelt districts from Minnipa to Cowell.

DISTINGUISHING FEATURES ➤ Distinctive, crowded, decussate leaves, coupled with white or cream, lateral flower clusters.

CULTIVATION ➤ Virtually unknown in cultivation but should adapt well to semi-dry or temperate areas on most soils, including limestone.

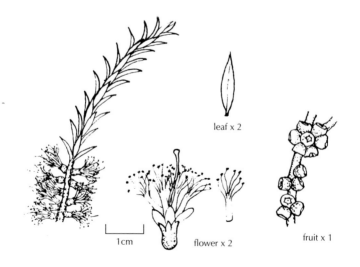

Worth cultivating, particularly in semi-arid gardens.

bushy shrub to 2m

M. pallescens Byrnes

DESCRIPTION ➤ Woody shrub, to 4m high, with hard, furrowed bark.
LEAVES tiny, only 1–4mm long, glabrous, ovate to narrowly ovate, peltate, spirally arranged, tip normally sharply pointed.
FLOWERS pale pink to pale mauve with reddish petals, in spikes mainly 15–20mm long; flowers in triads. Flowering season: spring.
FRUITS urn-shaped to cupular, 4–6mm long and wide.

DISTRIBUTION ➤ Western Darling Downs of Qld, from Miles district to Inglewood, favouring heavy clay-pan soils.

DISTINGUISHING FEATURES ➤ Tiny, ovate, peltate leaves coupled with hard, furrowed bark and flowers in triads on the spikes. The leaves of *M. tamariscina* (p. 150) and *M. irbyana* (p. 150) are somewhat similar, but these species feature papery bark.

CULTIVATION ➤ Often flowers quite profusely and, with its foliage tending to grow to near ground level, can be used as an ornamental screening or background shrub. Used in Qld for median strip planting and well suited to sub-tropical areas with well-drained soils. Would probably grow well in warm temperate areas but evidence is minimal.

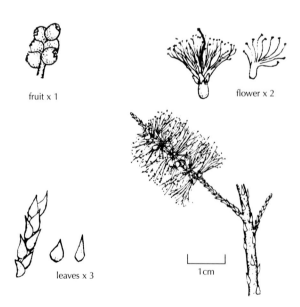

fruit x 1

flower x 2

leaves x 3

1cm

A useful shrub for heavy soils.

woody shrub

M. pauperiflora F. Muell. subsp. *mutica* Barlow

DESCRIPTION ➤ Dense, large shrub or small tree, usually 3–5m, with dark green foliage and rough, grey bark.
LEAVES spirally arranged, glabrous or glabrescent, except on new shoots; small, usually 3–6mm long by 1mm wide; thick, flat to terete, obtuse or acute but not pungent, and shortly petiolate.
FLOWERS profuse, white or cream, in small single or multiple rounded heads about 18mm wide, subtended by ovate bracts; hypanthia mostly glabrous but inflorescence axis tomentose. Flowering season: normally September–November.
FRUITS barrel-shaped, 3–4mm diameter by 4–5mm long, occurring singly or in small clusters.

DISTRIBUTION ➤ Widespread in SA, particularly on the Eyre Peninsula, and occurring from Nullarbor Plain to Murray Bridge in mallee woodland.

DISTINGUISHING FEATURES ➤ Small, thick, non-prickly leaves, rough, dark, grey bark, and small rounded flowerheads in monads (3–10).

M. pauperiflora F. Muell. subsp. ***pauperiflora*** from southern WA differs in its longer leaves (up to 13mm) with a pungent mucro and glabrous inflorescence axis.
  M. pauperiflora subsp. ***fastigiata*** Barlow, which is generally found further inland in WA, is similar to subsp. *pauperiflora* but can be distinguished by its broom-like habit and tomentose inflorescence axis.

SIMILAR SPECIES ➤ *M. sheathiana* W.V. Fitzg., Boree, from the dry goldfields of WA and further east, differs in its smaller, narrowly spathulate, sub-terete leaves, hairy hypanthium and generally, smaller inflorescences (to 14mm wide).

CULTIVATION ➤ All useful shrubs for winter-rainfall dry and temperate climates in most soils, including alkaline, provided drainage is reasonable. Good for hedging or windbreaks. Frost hardy.

large, dense shrub

Excellent plants for limestone and a semi-dry climate.

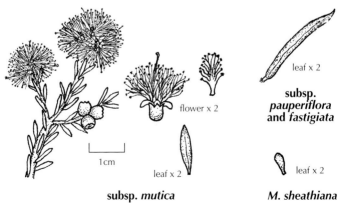

subsp. *mutica*

subsp. *pauperiflora* and *fastigiata*

M. sheathiana

M. pentagona var. *latifolia* Benth.

DESCRIPTION ➤ Normally an erect, medium-sized shrub, 1–1.5m high, but very old specimens are known which are almost small tree size.
LEAVES deep green, flat, spirally arranged, mainly 10–18mm long to 2–3mm wide, glabrescent except for silky-pubescent new growth, mainly narrowly elliptic to oblong, with a sharp tip, well spaced along branches.
FLOWERS bright pink, usually profuse, in terminal and axillary heads to 20mm wide; hypanthia glabrous, or sometimes with a few scattered hairs near the base; 5 short, distinct calyx lobes. Flowering season: October–early November
FRUITS form 'soccer-ball' clusters, 18–20mm diameter, calyx lobes weathering away.

DISTRIBUTION ➤ Eyre district of WA, favouring sandy coastal locations from just west of Esperance to Israelite Bay. Common at Quagi Beach.

DISTINGUISHING FEATURES ➤ Flat, narrowly elliptic to oblong or narrowly obovate, deep green leaves combined with 'soccer-ball' fruiting clusters and profuse pink flowers, with glabrous hypanthia.

SIMILAR SPECIES ➤ *M. caeca* Craven, from Arrino–Gin Gin Brook districts of WA, is a smaller shrub with similarities to *M. pentagona* var. *latifolia*. It features pink inflorescences to 15mm wide, flat, narrowly obovate leaves to 20mm long, a hairy hypanthium, deciduous petals and 15–20 ovules per locule (cf. 5–8 in the M. pentagona group).

CULTIVATION ➤ *M. pentagona* var. *latifolia* has been frequently cultivated, proving adaptable to a range of soils and conditions in dry or temperate areas. Should be tried in coastal gardens. Seedlings sometimes take many years to flower.
 M. caeca is rare in cultivation but is growing well at Victor Harbor, SA, in acidic sandy loam.

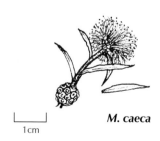

M. caeca

1cm

A very showy pink-flowering shrub.

M. caeca

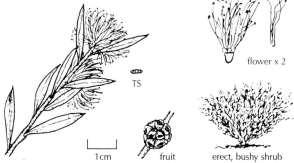

flower x 2

1cm fruit erect, bushy shrub

M. pentagona Labill. var. *pentagona*

DESCRIPTION ➤ Small to large, spreading or erect, prickly shrub, 1–3m high by 2–4m wide.
LEAVES spirally arranged, glabrescent, grooved, terete and pungent, usually 8–15mm long but variable.
FLOWERS in dense, globular heads to 20mm wide; pink to purplish; 2–5 stamens per claw; hypanthia glabrous, 1–1.8mm long; petals caducous. Flowering season: October–November.
FRUITS clustered into a honeycomb-like sphere, about 20mm diameter.

DISTRIBUTION ➤ WA, from Esperance westwards along the coast to Albany–Mt Barker district.

DISTINGUISHING FEATURES ➤ Distinctly spherical, honeycombed fruiting clusters, pungent, needle-like leaves and profuse flowering, each flower with 2–5 stamens per claw and hypanthium 1–1.6mm long.

The variety ***raggedensis*** Craven, from the western end of Mt Ragged, WA, differs by more stamens per claw (4–7), a longer hypanthium (1.5–2mm long) and usually more ovules per locule.

SIMILAR SPECIES ➤ *M. carrii* Craven, widely distributed from Eneabba to Esperance in WA, is somewhat similar but it is a smaller shrub which has been confused with *M. pentagona*. It differs by its deciduous petals and ungrooved leaves.

CULTIVATION ➤ *M. pentagona* has proved reliable under most conditions in temperate Australia where rainfall exceeds about 400mm annually, forming a magnificent shrub when in full flower.
  M. carrii is virtually unknown in cultivation.

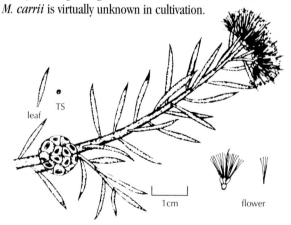

M. carrii

M. pentagona

M. pentagona is a magnificent, adaptable shrub for temperate regions.

M. carrii

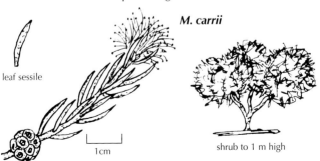

leaf sessile

1cm

shrub to 1 m high

M. platycalyx Diels, in Diels & Pritzel

DESCRIPTION ➤ Fairly small, woody or twiggy, glabrous, spreading shrub, 0.5–1.5m high and across.

LEAVES decussate, bluish green, ovate-elliptic, 6–15mm long by 2–7mm wide, obtuse or acute, with prominent marginal and central veins and oil glands.

FLOWERS pinkish purple to mauve, in opposite pairs or groups of 2–4 monads, along lateral branches; petals reflexed; hypanthium has very flat base; stamens 25–36 per bundle. Flowering season: spring.

FRUITS 6mm long by 8mm across, thickened at the base and fused into branch in opposite pairs at right angles.

DISTRIBUTION ➤ Inland WA, often growing on shallow soils, occurring from Latham district south and eastwards to Ongerup and Lake King districts.

DISTINGUISHING FEATURES ➤ Ovate-elliptic, bluish green leaves in opposite pairs, coupled with pinkish to mauve, single, sessile flowers in monads, fruits in opposite pairs fused onto branch.

SIMILAR SPECIES ➤ *M. adenostyla* Cowley, from near Dumbleyung eastwards to Hyden–Newdegate district of WA, is a larger, related shrub, differing by its cream flowers with more per inflorescence, glandular style and fewer stamens per bundle (11–21), narrower leaves and other minor features.

CULTIVATION ➤ Neither species is well known in cultivation, although *M. platycalyx* has occasionally been grown in light, well-drained soils in Adelaide, Melbourne and WA, usually forming a very small shrub with attractive flowers. Appears unsuited to more humid areas.

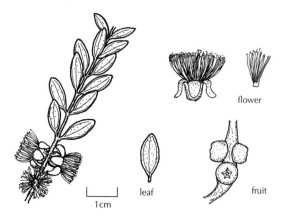

An attractive dense shrub for foreground planting.

small, woody shrub

M. plumea Craven

DESCRIPTION ➤ Small, rounded, bushy, ground-hugging shrub, normally about 1m high and wide, but may grow to 2m.

LEAVES spirally arranged, fleshy, 5–10mm long by 1–2mm wide, clavate or oblong, the apex obtuse or rounded or shortly acuminate; older leaves glabrescent, younger ones softly pubescent; transverse section may be elliptic, sub-circular or flattened-obovate.

FLOWERS profuse, deep pink to purple, in single or multiple heads; buds and brown bracts subtending heads covered with white, woolly or fluffy hairs, as are hypanthia and lobes; stamens 5–8 per bundle. Flowering season: September–November.

FRUITS urceolate, about 4mm long, in small non-globose clusters, the lobes weathering away.

DISTRIBUTION ➤ Salmon Gums, Scaddan to near Esperance districts of WA.

DISTINGUISHING FEATURES ➤ Masses of globular, pink flowerheads with white, fluffy buds, bracts, hypanthia and lobes.

SIMILAR SPECIES ➤ *M. rigidifolia* Turcz. ranges widely from Stirling Range–Albany district to Lake Cronin–Esperance–Wittenoom Hills district. This species has similarities to *M. plumea* but the absence of fluffy hairs on the floral parts distinguishes it, as well as its usually globose fruiting habit. Flowers prolifically in spring.

CULTIVATION ➤ *M. plumea* has occasionally been cultivated with success in Adelaide, in deep acidic sand and neutral clay, flowering early from seedlings.

The author has no knowledge of *M. rigidifolia* in cultivation.

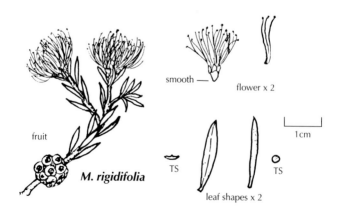

M. rigidifolia

M. rigidifolia

M. plumea

Profuse pink-flowering shrubs for the garden.

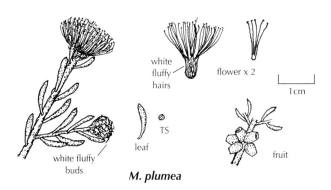

M. plumea

M. podiocarpa Barlow ex Craven

DESCRIPTION ➤ Spreading, prickly shrub, normally 1–2.5m high and often wider, with curled, flaky bark.

LEAVES alternate, glabrescent to glabrous, ovate-lanceolate or narrowly ovate, acuminate and pungent, 5–14mm long.

FLOWERS white, occurring mainly within shrub, laterally along older branches, but also from tips of young shoots; inflorescence comprises 1–3 monads; hypanthia hairy, 3–4mm long; stamens 30–45 per bundle. Flowering season: October to January.

FRUITS solitary, about 6mm long and wide, more or less cup-shaped, sepals remaining.

DISTRIBUTION ➤ Lake King to Grass Patch districts of WA, occurring in a range of habitats and soils from heavy clay to sand.

DISTINGUISHING FEATURES ➤ Pungent-pointed, ovate-lanceolate to narrowly ovate leaves, white flower clusters mainly hidden within the foliage, in monads of 1–3.

SIMILAR SPECIES ➤ *M. cliffortioides* Diels is found from Ravensthorpe district to Norseman in WA on a variety of soils and habitats. It is a similar shrub, easily separated by its much fewer stamens per bundle (8–13). Profuse in flower at its best. Flowering season: spring.

CULTIVATION ➤ Rarely cultivated (although *M. cliffortioides* was once well established in the Adelaide Botanic Garden), but both species should be very adaptable in semi-dry to temperate climates on a range of soils from heavy clay to sand, acidic or alkaline.

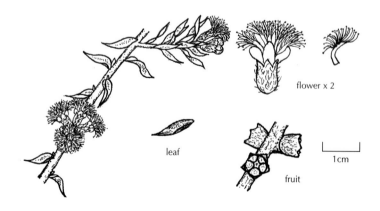

flower x 2

leaf

fruit

1cm

A shrub for most soils in temperate or semi-arid areas.

spreading shrub

M. polycephala Benth.

DESCRIPTION ➤ Twiggy, tangly, bushy shrub about 1m high and wide.
LEAVES spirally arranged, very reflexed or spreading, 4–15mm long by 2–4mm wide, ovate-elliptic to elliptic, more or less glabrous, sessile, sharp-pointed with prominent central and marginal veins on upper surface.
FLOWERS dusky pink, in small globular heads only 10–14mm wide; stamens 3 per bundle. Flowering may be quite sparse, or sometimes profuse. Flowering season: usually September–October.
FRUITS tiny capsules forming small, globose cluster about 5–6mm in diameter, lobes weathering away.

DISTRIBUTION ➤ Fairly rare species from Gnowangerup–Pingrup–Jerramungup district of WA, mostly in sand.

DISTINGUISHING FEATURES ➤ Twiggy, tangly shrub with small, spreading, widely spaced, ovate-elliptic to elliptic leaves tapering to a sharp point.

SPECIES WITH SIMILAR ELLIPTICAL LEAVES ➤ *M. pauciflora* Turcz. is another shrubby species from WA, from the swampy south-west Perth area to Augusta and Albany districts as far as Cape Riche. It has smooth but decussate, narrowly elliptic leaves, 3–12mm long by 1–2mm wide. Tiny heads of white flowers in the laterals comprise 2–8 monads and 2–7 stamens per bundle. It is an unspectacular shrub.

CULTIVATION ➤ *M. polycephala* is not often seen in cultivation but is claimed to be an adaptable, showy species in well-drained soils with an annual rainfall exceeding about 350mm. It is intolerant of humidity. Cultivated specimens seen by the author have been disappointing because of the shrub's habit and lack of flowers (see picture opposite).

   M. pauciflora has little garden potential.

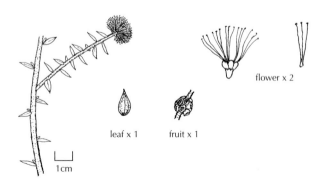

leaf x 1 fruit x 1 flower x 2

1cm

Mainly a shrub for the enthusiastic collector.

M. pauciflora

bushy, tangly shrub
(*M. polycephala*)

M. preissiana Schauer, in Lehm.

DESCRIPTION ➤ Normally a bushy-crowned paperbark tree, 6–10m, but sometimes a shrub.
LEAVES narrowly elliptic to narrowly ovate, glabrescent but pubescent when young, fairly stiff and spreading.
FLOWERS white or cream, in cylindrical spikes 30–50mm long by 20mm wide, singly or several together; stamens pinnately arranged (27–30 per bundle); petals reflexed and spathulate; axis growing on prominently. Flowering season: spasmodic, usually between August and March.
FRUITS about 4mm long and wide, barrel-shaped, loosely clustered along branches.

DISTRIBUTION ➤ Low-lying, wet areas of south-west of WA, from near Eneabba to east of Albany, including metropolitan Perth.

DISTINGUISHING FEATURES ➤ Narrow, more or less spreading leaves; tree dimensions and papery bark; white flower spikes, normally with spathulate petals and many stamens per bundle.

CULTIVATION ➤ Good, small, paperbark tree for wet, swampy, non-saline soils. Suited to temperate areas where rainfall exceeds about 400mm annually, but rarely seen in cultivation.

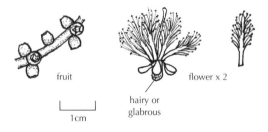

fruit

1cm

hairy or glabrous

flower x 2

A handsome paperbark for wet soils.

M. procera Craven

DESCRIPTION ➤ Fairly loosely-branched, slender, upright shrub to 1–2m.

LEAVES spirally arranged, 6–19mm long by 1–1.5m wide, glabrous, linear to linear-obovate, more or less sub-circular in transverse section, apex obtusely acuminate.

FLOWERS light pink to pinkish mauve, arranged in triads, in heads to 30mm wide; hypanthia hairy; calyx lobes scarious throughout, scarious in a marginal band, or herbaceous to margin; petals caducous; stamens 8–10 per bundle. Flowering season: normally late spring or December.

FRUITS irregularly peg-like, 3–4mm long, calyx lobes weathering away.

DISTRIBUTION ➤ Kulin–Karlgarin–Lake Grace district of WA.

DISTINGUISHING FEATURES ➤ Smooth leaves as described, slender, erect habit and pink flowerheads to 30mm wide.

SIMILAR SPECIES ➤ *M. villosisepala* Craven, from Southern Cross–Coolgardie district south to Stirling Range–Ravensthorpe district of WA, also features pinkish flowerheads in triads, but smaller, to 20mm wide. Leaves oblong to linear, 3–15mm long, silky-hairy to glabrescent. Fruits are peg-like. Flowering season: any time from September well into summer.

M. wonganensis Craven, from Wongan Hills district of WA, is very similar to *M. villosisepala*, differing in its normally dimorphic leaf hairs. Flowering heads are purple to deep mauve to 25mm wide, also arranged in triads. Flowering season: September–October.

CULTIVATION ➤ *M. procera* is growing well in deep, acidic, white sand near Adelaide but the author has no other knowledge of these three species in cultivation.

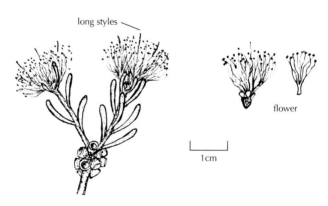

This erect, slender shrub is ideal for a narrow garden spot.

spindly erect shrub

M. psammophila Diels, in Diels & Pritzel

DESCRIPTION ➤ Small, bushy shrub, 0.5–1m or more high.
LEAVES 2–8mm long by 0.5–1.5mm wide, crowded and spirally arranged; blade features long, sparse, pubescent hairs, contrasting with dense, woollier hairs on branchlets; oil glands prominent; apex obtuse or rounded; 2 shallow channels on undersurface.
FLOWERS in small, reddish pink heads; yellow anthers conspicuous and style long; heads to 14–25mm wide, in triads (1–4); hypanthia hairy with very pointed acuminate lobes so that buds often appear rostrate; stamens 7–15 per bundle. Flowering season: September–December.
FRUITS distinctly urn-shaped, in non-globose or peg-like clusters.

DISTRIBUTION ➤ Kalbarri–Geraldton districts of WA, in sand.

DISTINGUISHING FEATURES ➤ Beaked (rostrate) buds; small, sparsely hairy, obtuse, channelled leaves and distinctly urn-shaped fruits.

SIMILAR SPECIES ➤ *M. bisulcata* F. Muell., also from Kalbarri district of WA, is very close in most features (flower colour, leaf shape and size, etc.) but differs in not having long, acuminate or beaked calyx lobes.

CULTIVATION ➤ *M. psammophila* has been successfully cultivated in temperate areas (Adelaide, Perth, Melbourne), in soils which are light and well drained, acidic to neutral, in warm sunny locations. Profuse flowering with dense groundcover.

The author has no knowledge of *M. bisulcata* in cultivation, but both species would form attractive, small foreground or rock garden shrubs.

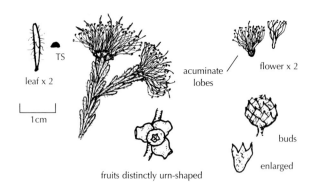

leaf x 2
TS
1cm
acuminate lobes
flower x 2
fruits distinctly urn-shaped
buds enlarged

A spectacular, ground-hugging shrub.

small shrub to 1 m

M. pulchella R. Br., in Ait.
Claw Honey-myrtle

DESCRIPTION ➤ Small, dainty shrub, usually about 0.5m high and spreading, with many arching branches.
LEAVES spirally arranged, small and crowded, obovate to elliptic, 2–6mm long by 1–3mm wide, with prominent oil glands on undersurface.
FLOWERS pink to pinkish mauve, with violet sepals; solitary, often in opposite pairs; claw-like, with claws curving into the centre of flower. Flowering season: appear over long periods from October to autumn.
FRUITS urn-shaped, about 6mm across, with sepaline teeth.

DISTRIBUTION ➤ Southern WA, from Hopetoun east to Israelite Bay, in sand heath which is often waterlogged in winter, and a variety of other habitats.

DISTINGUISHING FEATURES ➤ Solitary, claw-like flowers featuring numerous short stamens at the base and a few curving ones at the end of each staminal claw.

CULTIVATION ➤ Popular, long-flowering, small shrub. Has proved reliable in a wide range of soils in winter-wet temperate areas. Good dwarf plant for bordering a garden lawn as it covers the ground well and tolerates the additional water received, provided the soil has reasonable drainage. Could be used as a container plant with judicious pruning. Probably resents highly alkaline soils.

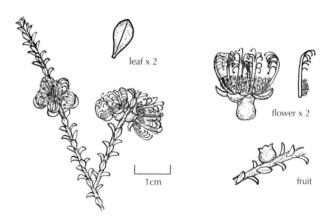

This popular, dainty shrub has unusual flowers.

low, spreading shrub

M. pungens Schauer, in Lehm.

DESCRIPTION ➤ Low, spreading, very prickly shrub, 0.5–1m high and wide.

LEAVES alternate, sessile, narrow, more or less linear, 10–35mm long by about 1mm wide, splayed in all directions, closely hairy and very prickly due to a small, sharp mucro; young growth silky.

FLOWERS in rounded or elongated heads about 15mm wide, bright yellow and numerous; hypanthia closely hairy. Flowering season: September–October.

FRUITS 3–4mm wide, in fused clusters, raised valves pubescent on both surfaces.

DISTRIBUTION ➤ Fairly widespread in southern districts of WA, from Stirling, Avon and Eyre botanical districts to southern edge of Coolgardie district.

DISTINGUISHING FEATURES ➤ Very pungent, splayed, linear, stalkless leaves coupled with numerous yellow inflorescences.

CULTIVATION ➤ Adapts well to temperate areas with well-drained acidic to neutral soils, but is unknown to the author in other less favourable conditions. Frost hardy. Cuttings strike readily. Its very prickly foliage would be excellent for controlling foot traffic.

This species may be removed from *Melaleuca*, as it is not entirely consistent with the criteria for the genus.

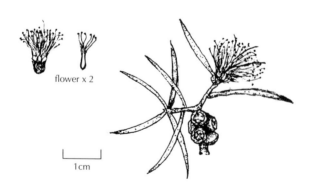

flower x 2

1cm

This floriferous low shrub is exceptionally prickly.

low, spreading shrub

M. pustulata J.D. Hook.

DESCRIPTION ➤ Medium to large, bushy shrub, 2–4m high and wide.
LEAVES glabrescent, spirally arranged, linear-elliptic, obtuse in side view, club-shaped at tip, pustulate; young growth hairy.
FLOWERS profuse, pale lemon, strongly scented; in small, ovoid or globular, terminal spikes; concave brown bracts surround small, smooth hypanthium. Flowering season: October.
FRUITS usually 4–5mm long and wide, cup-shaped, scattered or in crowded spikes.

DISTRIBUTION ➤ Endemic to Tasmania's central and east coast regions in wet places.

DISTINGUISHING FEATURES ➤ Thick, pustulate leaves with club-shaped warty tips in side view, and pale lemon inflorescences.

CULTIVATION ➤ Apparently does well in acidic, well-drained but winter-wet soils in Vic. and Tas., where it is considered a good hedging or windbreak plant for coastal gardens. Unlikely to tolerate exposure to harsh salt-laden winds in SA and WA.

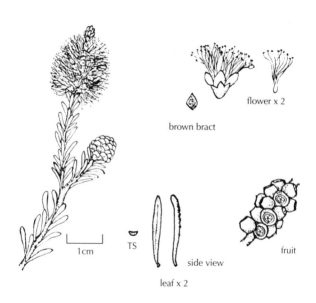

An attractive large and bushy shrub for winter-wet soils.

medium to large, bushy shrub

M. quadrifaria F. Muell.

DESCRIPTION ➤ Large, small-leaved shrub, 2–5m, with dark fibrous bark.

LEAVES decussate, oblong or narrowly ovate to narrowly elliptic, 2–7mm long, sessile, fleshy, often curving upwards.

FLOWERS white or cream, in small spikes of 2–9 triads; hypanthia glabrous or faintly pubescent. Flowering season: spring and summer.

FRUITS small capsules, 3–4mm diameter, cup-shaped or spherical, in small irregular clusters.

DISTRIBUTION ➤ WA, on the Fraser Range east of Norseman, on Mt Ragged inland from Israelite Bay, and Newdegate area to Balladonia. Also occurs in western SA near Eucla.

DISTINGUISHING FEATURES ➤ Small, up-curved, decussate leaves.

CULTIVATION ➤ Useful shrub in cultivation because of its success in sheet limestone soils and an arid climate. Successful in most well-drained soils where annual rainfall is less than about 700mm and the climate is dry to temperate.

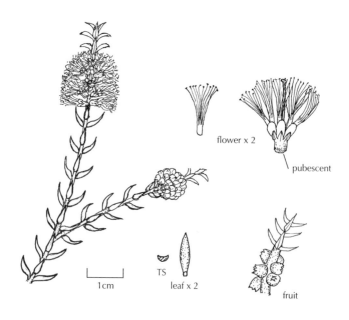

A shrub suited to arid conditions and limy soils.

large shrub

M. quinquenervia (Cav.) S.T. Blake
Broad-leaved Paperbark

DESCRIPTION ➤ Erect, small to medium-sized tree, normally 8–12m, but sometimes reaching 25m, with thick, white or greyish, papery bark.
LEAVES flat, stiff and leathery, 40–120mm long by 10– 30mm wide, lanceolate-elliptic, narrowly elliptic or oblanceolate, with 5–7 longitudinal nerves.
FLOWERS white, cream or greenish, in bottlebrush-like spikes 20–50mm long; stamens 4–10 per bundle. Flowering season: autumn–winter.
FRUITS small, woody capsules about 5mm diameter, flat-topped, clustered along branches in spike-like formation

DISTRIBUTION ➤ One of the most common paperbark trees of Australia's east coast, mainly along streams and in swamps. It extends from near Sydney north to Cape York, usually within 40km of the coast, and is more common in southern half of its range. Also in New Guinea, Indonesia and New Caledonia.

DISTINGUISHING FEATURES ➤ Broad, flat, stiff, leathery, 5–7 nerved leaves, usually lanceolate-elliptic, coupled with broad, cylindrical flower spikes which occur in autumn and winter.

CULTIVATION ➤ Very adaptable tree, easily grown under most conditions from Adelaide to the tropical north. Probably requires a rainfall in excess of 450mm annually. Frequently seen as a street tree in Sydney. Has colonised to pest proportions in Florida, USA.

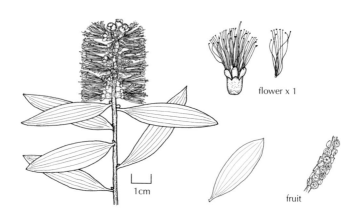

A very adaptable tree for tropical to temperate areas.

erect tree

M. radula Lindl.
Graceful Honey-myrtle

DESCRIPTION ➤ Open, woody, erect or spreading, shrub, 1–5m.
LEAVES decussate, linear or narrowly elliptic, acute to acuminate, 10–50mm long by 1–3mm broad, concave or with incurved margins, with prominent oil glands on undersurface.
FLOWERS in sessile, opposite pairs, normally forming a spike of monads about 40mm long, but sometimes reduced to 1 or 2 pairs on branch; delicate mauve, violet, pink or white, and rather fluffy, due to the numerous stamens on each claw (30–90); stigma broad and green. Flowering season: mainly July–September.
FRUITS smooth and globular, 6–8mm in diameter.

DISTRIBUTION ➤ Widespread in WA, in Swan River area, in central wheatbelt north to Kalbarri district and east to Norseman area, in a variety of habitats and often on alkaline sands.

DISTINGUISHING FEATURES ➤ Smooth, globular fruits, irregular, normally mauve to pink, fluffy flower spikes, the flowers in loose opposite pairs. This species has floral similarities with *M. fulgens* (p. 114), with garden hybrids known. The rounded fruits of *M. radula*, however, are distinctive.

CULTIVATION ➤ Adaptable shrub for winter-rainfall semi-dry to temperate areas, and suited to most soils with good drainage. Good forms are very showy in full flower. Prune regularly to avoid a straggly bush and maintain profuse flowering.

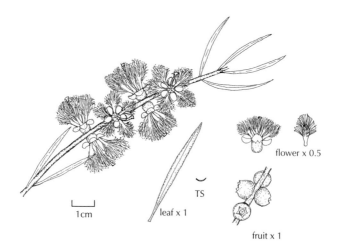

This is a medium to large, profusely flowering garden shrub.

open, woody shrub

M. radula hybrid
(possibly *M. radula* x *M. fulgens*)

DESCRIPTION ➤ Shrub, normally 1–3m high, with many slender, ascending branches.
LEAVES decussate, glabrous, linear, concave, acute, with prominent glands on undersurface, mostly 10–30mm long by 1–2mm wide.
FLOWERS deep purple to purplish pink with yellow anthers, in cylindrical, lateral spikes 30–40mm long and wide. Flowering season: long period in spring, spasmodically at other times, especially summer.
FRUITS flattened-globular, 10mm or more in diameter, but variable, sometimes urn-shaped like *M. fulgens*.

DISTRIBUTION ➤ Garden origin.

DISTINGUISHING FEATURES ➤ Gold-tipped purplish flower spikes resembling *M. fulgens*, but generally smaller, similar in size to *M. radula*.

CULTIVATION ➤ Adaptable shrub which grows in most soil types, flowers well in full sun or semi-shade, but becomes chlorotic in strongly alkaline soils. Only known in temperate plantings, but *M. fulgens* is grown in the Brisbane area, suggesting this hybrid may also succeed on the east coast.

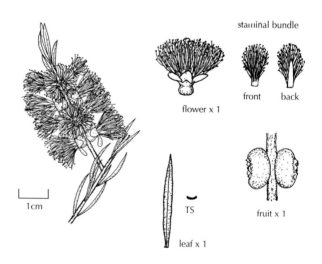

Woody or twiggy shrub with beautiful flowers.

erect shrub

M. rhaphiophylla Schauer, in Lehm.
Swamp Paperbark

DESCRIPTION ➤ Usually a bushy-crowned tree, 7–10m high, or a large shrub, with greyish white, papery bark.

LEAVES terete to narrow-linear, grey–green to green, glabrescent except on pubescent young shoots, 10–50mm long, tapering to a slight hooked point when terete.

FLOWERS white to cream, in loosely arranged ovoid spikes 25–35mm wide by 20–40mm long; profuse and showy in good forms. Flowering season: normally November, but spasmodic.

FRUITS cylindrical to cup-shaped, 5–6mm diameter, in sparse clusters.

DISTRIBUTION ➤ Widely distributed in south-west of WA, along river banks and in swamps from Kalbarri to Albany district.

DISTINGUISHING FEATURES ➤ Normally grey–green, slightly hooked, needle-like or flat linear leaves, and larger than average ovoid, white to cream flower spikes, usually prominently displayed.

CULTIVATION ➤ Good small landscape tree for lining permanently wet sites such as reservoirs, lakes and man-made wetlands, showy in flower. Suited to low-saline wet sites in temperate areas, where it has been used on occasions. No evidence in other conditions.

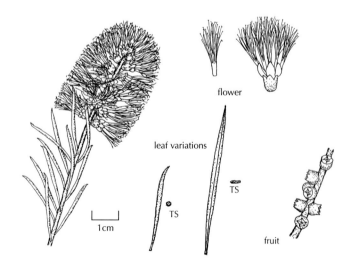

A showy tree for wet sites.

bushy-crowned tree

M. ringens Barlow

DESCRIPTION ➤ Low, dense, spreading shrub, usually under 1m high and about 2m wide, but taller and thinner in semi-shade.
LEAVES glabrescent except for hairy young shoots, narrowly ovate to elliptic, spirally arranged, crowded and reflexed, apex acute to obtuse, slightly petiolate.
FLOWERS profuse, cream with yellow anthers on spikes 20–60mm long by 20mm wide, the axis growing on; individual flowers tightly packed on spike; sickly smelling but showy. Flowering season: September–October.
FRUITS woody, bell-shaped capsules with thickened, spreading, sepaline teeth; about 5mm long and wide, clustered into a loose spike.

DISTRIBUTION ➤ Point D'Entrecasteaux (Windy Harbour) in WA, growing in sand over limestone on exposed coastal cliffs and ridges, and in Albany district.

DISTINGUISHING FEATURES ➤ Bright green foliage, similar to that of *M. diosmifolia* (p. 96), but with smaller leaves, showy creamy yellow flower spikes, and usually low, spreading habit.

CULTIVATION ➤ This shrub has crept into cultivation as a dwarf form of *M. diosmifolia*. It is adaptable to most soils in temperate areas, including limestone, where rainfall is above 500mm annually.

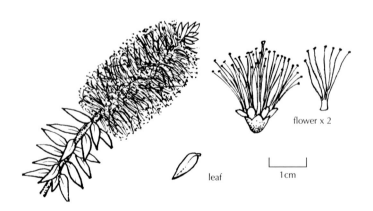

This dense, spreading shrub is showy in flower.

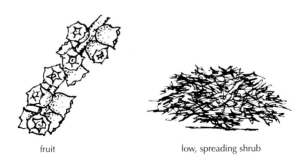

fruit

low, spreading shrub

M. ryeae Craven

DESCRIPTION ➤ Smallish shrub to 1.5m high, but usually less, often spreading and semi-prostrate.
LEAVES small, spirally arranged, generally 5–8mm long, obovate to elliptic, obtuse and 3–veined. New growth woolly-hairy.
FLOWERS profuse deep pink to purplish pink, in dense, globular heads to 25mm wide; hypanthia covered in white hairs; calyx lobes scarious throughout; petals usually deciduous on opening. Flowering season: normally October.
FRUITS small capsules about 3mm long by 4mm across, usually in globose or rounded clusters.

DISTRIBUTION ➤ Northern sand plains of WA, from Arrowsmith River south to Bullsbrook.

DISTINGUISHING FEATURES ➤ Small, obovate to elliptic leaves, low habit, profuse, pink to purplish pink flowerheads and normally globose fruiting clusters, which may also be irregularly peg-like, causing identification from *M. amydra* to be difficult.

SIMILAR SPECIES ➤ *M. amydra* Craven, from Arrowsmith River district to Dandaragen–Moora district of WA, is almost identical, differing in its narrower, mostly elliptic leaves and irregular, peg-like fruiting clusters.

CULTIVATION ➤ Both species (as *M. leptospermoides*) have been cultivated with success in Adelaide and other winter-rainfall temperate areas on a range of soils. They form attractive, compact shrubs which are particularly showy in flower.

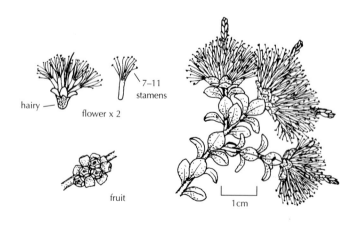

Adaptable, and one of the best pinks for the garden.

low, compact shrub

M. sapientes Craven

DESCRIPTION ➤ Dense, spreading shrub, usually to 1m, but sometimes taller, with silvery grey, shimmering foliage.
LEAVES spirally arranged, linear, usually curved, normally 10–15mm long, densely covered with soft, silky hairs; blade in transverse section usually sub-lunate to shallowly lunate, linear, or sometimes semi-elliptic.
FLOWERS pink or pinkish mauve; in small, terminal heads; hypanthia and calyx lobes silky-hairy; style 7.5–8mm long; stamens 6–10 per bundle. Flowering season: normally late spring or early summer.
FRUITS 3–4mm diameter, arranged in small, irregular, peg-like cluster.

DISTRIBUTION ➤ WA, from the Hyden–Jerramungup area to Salmon Gums–Ponier Rock district.

DISTINGUISHING FEATURES ➤ Silky or hoary, soft silvery grey foliage.

SIMILAR SPECIES ➤ *M. holosericea* Schauer in Lehm., from the Toodyay–Northam districts of WA is a rare small shrub with similar foliage. Never seen by the author.
   M. idana Craven, from the Kalbarri–Wannoo district of WA, has similar foliage but differs in transverse section of leaf (quadrate to broadly oblong or semi-circular), and longer styles (11–13mm) on the attractive pink flowerheads.

CULTIVATION ➤ *M. sapientes* has long been in cultivation (as *M. holosericea*) because of its beautiful foliage. Has adapted well to most soils.

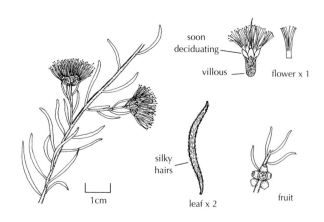

This is a foliage shrub of great beauty that makes a superb garden feature.

low, silky shrub

M. scabra R. Br., in Ait.
Rough Honey-myrtle

DESCRIPTION ➤ Dwarf to small shrub, about 30cm–1m high and wide.
LEAVES spirally arranged, linear, linear-obovate or falcate, 5–20mm long by about 1mm or more wide, glabrous and warty, apex pointed to obtuse; often channelled on underside.
FLOWERS profuse, in dense rounded heads to 22mm wide, comprising 1–5 triads; filaments deep pink to magenta, anthers yellow; hypanthia glabrous or hairy; petals deciduous; stamens 3–7 per bundle; style 6–10mm long. Flowering season: July–November.
FRUITS in globose clusters, each capsule 3–4mm long, lobes weathering away.

DISTRIBUTION ➤ South coast of WA, from Hopetoun to Israelite Bay and north to Mt Ragged.
   Prior to the *Melaleuca* revision of 1999, the name *M. scabra* was misapplied to many other species of similar habit and flowering, resulting in the belief that *M. scabra* was very widespread throughout the south-west.

DISTINGUISHING FEATURES ➤ Narrow, hairless but warty leaves, often channelled on underside; and hairless branchlets.

SIMILAR SPECIES ➤ ***M. papillosa*** Turcz. ex Craven, occurring in the Fitzgerald district of WA, is a similar shrub, mainly differing in its hairy leaves and silky-hairy floral axis, usually smaller flowerheads (to 18mm wide) and non-globose fruiting clusters, the lobes forming sepaline teeth. Flowering season: September–October.

CULTIVATION ➤ *M. scabra* is believed to have been cultivated over a long period, but many of these plants may well have been other species (e.g. *M. manglesii* and some forms of *M. trichophylla* have been grown as *M. scabra* in Adelaide since the 1960s), and thus it is not possible to give an opinion on either *M. scabra* or *M. papillosa* as cultivated plants.

A small, woody shrub with profuse pink flowerheads.

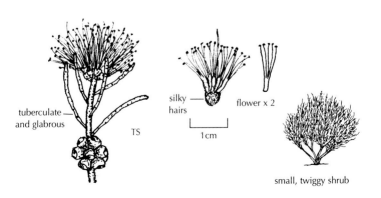

M. sclerophylla Diels, in Diels & Pritzel

DESCRIPTION ➤ Low, fairly dense shrub, usually under 1m tall and wide, but very old, much larger specimens can be seen in the wild.
LEAVES spirally arranged, often spreading, dark green and leathery, flat but rough due to the many tubercles on both surfaces; narrowly oblanceolate to narrowly elliptic, 8–30mm long by 2–6mm wide with prominent central vein; young growth light green, and soft due to a sparse covering of long woolly hairs, a few hairs persisting on older leaves.
FLOWERS in small, gold-tipped, purplish heads 15–20mm wide; rachis and hypanthia woolly-hairy; fragile white petals have ciliate margins; stamens 4–7 per bundle; buds enclosed within dark grey–green, ovate, deciduous bracts. Flowering season: usually early spring.
FRUITS in small, globular clusters of cup-shaped capsules about 2mm wide, lobes weathering away.

DISTRIBUTION ➤ WA, from Kalbarri district to Wongan Hills district in a range of soils and habitats, mostly in sandy or quartzite soils.

DISTINGUISHING FEATURES ➤ Rough, leathery, very tuberculate leaves as described, and profuse purplish flowerheads.

CULTIVATION ➤ Adaptable shrub which has succeeded in light and heavier clay soils with reasonable drainage, in areas of temperate Vic. and SA. Flowering period is quite short but showy in a good year.

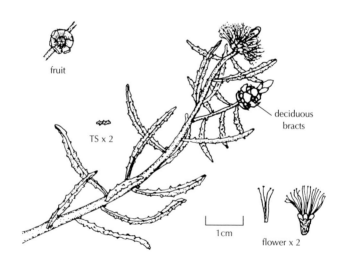

A useful smallish shrub for the garden.

small shrub

M. sieberi Schauer.

DESCRIPTION ➤ Large paperbark shrub or small tree, normally under 10m but sometimes to 20m.

LEAVES scattered and crowded, spirally arranged, mainly narrowly elliptic to elliptic, mostly 5–12mm long by 1–2mm wide, apex acute; glabrous or glabrescent when mature, young growth pubescent.

FLOWERS in profuse but small terminal spikes 10–20mm long, which may be in monads or triads on same branch; petals and stamens white or pinkish; buds and rachis woolly-hairy; stamens normally 11–20 per claw; hypanthia 2–3mm long. Flowering season: spring.

FRUITS cupular, 3–4mm long and wide, valves deeply inserted.

DISTRIBUTION ➤ Confined to Wallum heathland of south-east Qld and north-east NSW, in sandy soils with a high water-table.

DISTINGUISHING FEATURES ➤ Small, sharp-pointed, narrow, crowded and scattered leaves, and small white flower spikes with numerous stamens per flower, in monads or triads.

SIMILAR SPECIES ➤ ***M. deanei*** F. Muell., a rare and endangered NSW species from Sydney region to Nowra, has similar, but longer leaves (10–30mm) which feature pinnate to parallel-pinnate veins. It is a shrub 0.5–2.5m high with spikes of white flowers.

CULTIVATION ➤ *M. sieberi* is an adaptable species, at home in wet, sandy soils of the east coast. It suckers after fires and tolerates frost.
   M. deanei is not known to the author in cultivation.

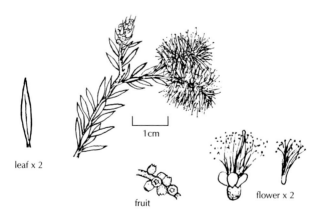

leaf x 2

1cm

fruit

flower x 2

An attractive small tree for gardens, especially in the sub-tropics.

small tree or large shrub

M. societatis Craven

DESCRIPTION ➤ Dwarf shrub, rarely more than 70cm high by 1m wide, but sometimes taller, to 2m.
LEAVES spirally arranged, short, fleshy, linear or oblong, more or less obtuse, mainly 3–8mm long by about 1mm wide, glabrescent.
FLOWERS profuse, deep pink, in small, globular heads only 8–15mm across, singly or several together, mostly along branches on short side branchlets; stamens 3–6 per bundle. Flowering season: normally October–November.
FRUITS tiny capsules in small, rounded or globose clusters.

DISTRIBUTION ➤ WA, from Salmon Gums to Israelite Bay, where it may be seen in places as the dominant understorey species in low woodland or shrubland. Also recorded for Stirling Range–Jerramungup district.

DISTINGUISHING FEATURES ➤ Short, normally obtuse, fleshy, narrow leaves and small, deep pink, globular flowerheads; mostly 3 stamens per bundle.

SIMILAR SPECIES ➤ *M. subtrigona* Schauer, in Lehm., a similar shrub from Brookton–Stirling Range–Ravensthorpe district of WA, has slightly larger inflorescences (up to 16m wide), quite warty leaves, due to the prominent oil glands, distinctly rounded calyx lobes and non-globose fruiting clusters, each capsule a flattened urn shape.

CULTIVATION ➤ Neither species is well known in cultivation although *M. societatis* has proved reliable in Adelaide's red–brown clays, where it has been grown against the butt of a large established eucalypt, similar to the way it is often seen in the wild as understorey to mallees. Flowering, however, has been sparse.

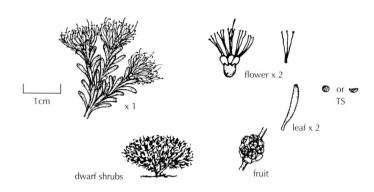

Both are attractive understorey shrubs.

M. subtrigona

M. subtrigona

leaf x 2
x 1
1cm
flower x 2 — rounded lobes
fruit

M. spathulata Schauer, in Lehm.

DESCRIPTION ➤ Normally a compact shrub, to about 1.5m high and wide, with many rigid, often twisting branches.

LEAVES sparsely arranged, fleshy, mainly obovate to spathulate, variable in size and proportions, but mainly 6–10mm long, pointed but not pungent; usually noticeably recurved, sometimes twisted, smooth; young growth soft and ciliate.

FLOWERS prolific, deep pink, terminal and lateral in single and multiple globular heads, usually 10–20mm across; hypanthia white-ciliate; staminal claw divides into 2–5 (usually 3) filaments. Flowering season: October–November.

FRUITS normally clustered into small heads resembling tiny pineapples, up to 1cm long.

DISTRIBUTION ➤ South coast and inland WA, from near Denmark to the Jerramungup area, including the Stirling Range.

DISTINGUISHING FEATURES ➤ Characteristic spiralling leaf arrangement on often divaricate branches, and slightly fleshy, recurved obovate to spathulate, pointed leaves.

CULTIVATION ➤ Very adaptable shrub, ornamental in full flower. Succeeds in light or heavy soils in areas of winter rainfall of 400mm or more and has done well in some areas of the more humid east coast. Frost hardy. Prune regularly to maintain a good shape and profuse annual flowering.

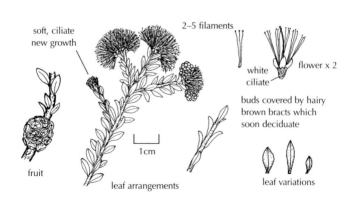

A beauty for the garden.

compact shrub

M. spicigera S. Moore

DESCRIPTION ➤ Branching shrub, normally 1–2m high and wide but often less, with rough, grey, fibrous bark.

LEAVES spirally arranged, blue–green, glabrous, ovate or cordate to almost oblong, sessile or faintly petiolate, often recurved, concave and undulate, 6–15mm long by 2–8mm wide, with prominent oil glands; young growth softly hairy.

FLOWERS in lateral cylindrical or ovoid spikes 10–20mm long and wide; petals and filaments pale pink to pinkish mauve; hypanthia and lobes white-hirsute; stamens 9–16 per bundle. Flowering season: normally September–October, sometimes later.

FRUITS tulip-shaped (due to the persistent sepals), 5–6mm long, in small irregular clusters.

DISTRIBUTION ➤ Inland WA, Minnivale district south to Ongerup district and eastwards to Salmon Gums district.

DISTINGUISHING FEATURES ➤ Ovate to oblong, glaucous, often wavy, more or less stem-clasping leaves, and pinkish flower spikes with broad stigmas, in lateral clusters.

CULTIVATION ➤ Good forms of this shrub are ornamental and well worth growing. It is suited to most well-drained soils in temperate or semi-dry climates within the 300–800mm rainfall range. Grows well in semi-shade or full sun and is frost tolerant.

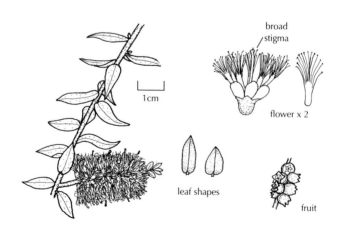

A useful garden shrub, particularly attractive in flower.

branching shrub to 2m

M. squamea Labill.
Swamp Honey-myrtle

DESCRIPTION ➤ An erect or bushy shrub, normally 1–3m but sometimes reaching 6m, with rough, grey bark.
LEAVES spirally arranged, crowded, narrowly ovate with 3 longitudinal nerves, acuminate, inflexed at the apex, normally 4–8mm long by 1.5–3mm wide; young shoots white-tomentose.
FLOWERS profuse, lilac to pinkish mauve, white in Tas. mountains; in terminal heads about 10mm long by 15mm wide; hypanthia silky-hairy; stamens 4–9 per bundle. Flowering season: spring.
FRUITS spherical but distinctly narrowed at the top, 6–7mm long and wide, in clusters.

DISTRIBUTION ➤ South-east Australia, occurring in Tas., Vic., NSW and SA in damp places. Grows in Royal National Park, Sydney.

DISTINGUISHING FEATURES ➤ Narrowly ovate, acuminate leaves, characteristically inflexed at the apex.

CULTIVATION ➤ Grows well in temperate and sub-tropical areas where there is ample moisture. Suited to non-saline, swampy soils which may dry out in summer. May fail in limy soils.

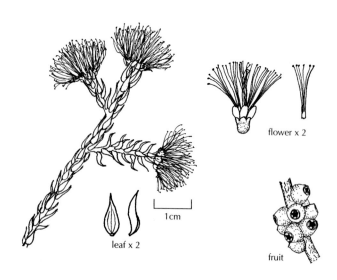

A good garden shrub for wet, non-saline soils.

erect or bushy shrub

M. squarrosa Donn. ex Smith
Scented Paperbark

DESCRIPTION ➤ Handsome, usually erect and dense, arching shrub, 2–6m, or sometimes a small tree, with greyish, papery bark. Prostrate in coastal situations.
LEAVES decussate, small and rigid, ovate to ovate-lanceolate, 5–10mm long by 5–7mm broad, acute, and distinctly 5–7 nerved.
FLOWERS cream, pleasantly perfumed; mainly in profuse terminal spikes 10–40mm long by about 15mm broad; each flower subtended by a leafy deciduous bract. Flowering season: normally spring or early summer.
FRUITS small, rounded or cup-shaped capsules, 3–4mm diameter, normally united into a spike.

DISTRIBUTION ➤ Widely distributed in south-eastern Australia, from south coast of NSW through Vic. (common in the Grampians), to SA (south-east and Kangaroo Island) and common in Tas., favouring low-lying, swampy locations.

DISTINGUISHING FEATURES ➤ Decussate, ovate leaves, 5–10mm long, with 5–7 distinct longitudinal nerves, and profuse, pleasantly scented, cylindrical, cream flower spikes.

CULTIVATION ➤ Excellent large shrub for winter-wet areas in a wide range of soil types. Also suited to the more humid east coast. Frost hardy. May resent very limy soils.

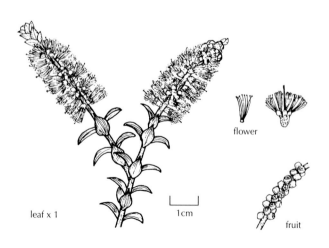

leaf x 1   1cm   flower   fruit

An excellent plant for winter-wet gardens.

tall, dense shrub

M. stramentosa Craven

DESCRIPTION ➤ Dense, bushy shrub, normally 0.5–1.5m high and wide.

LEAVES spirally arranged, sessile, fleshy, clothed in silky-woolly, matted hairs which gradually disappear as leaves age, 6–14mm long by 1–1.5mm wide, more or less terete (very narrowly obovate or oblong, but semi-circular or semi-elliptic in transverse section), the apex a sharp mucro, or sometimes obtuse.

FLOWERS profuse, bright pink to purplish, in small lateral and terminal heads 10–17mm across; hypanthia about 2mm long, covered in silky matted hairs; petals deciduous; stamens 4–6 per bundle. Flowering season: mainly October.

FRUITS cylindrical, about 4mm wide and long, occurring singly or in small, irregular or peg-like clusters, with sepaline teeth.

DISTRIBUTION ➤ Ravensthorpe district of WA, in heathland on clay or gravel.

DISTINGUISHING FEATURES ➤ Near-terete leaves clothed in silky or woolly, matted hairs becoming sparse as the leaves age, and small, pink–purple inflorescences with matted, hairy hypanthia and flower buds.

SIMILAR SPECIES ➤ *M. similis* Craven, from same area of WA, mostly in sand, is almost identical, differing in the non-matted silky hairs on the leaves and hypanthium, plus fruits where the calyx lobes weather away.

CULTIVATION ➤ *M. stramentosa* was once grown successfully by the author in slightly alkaline clay, but there is otherwise no evidence of either species being cultivated. Like most of these small West Australian melaleucas, however, they should adapt to most soils which are not too limy.

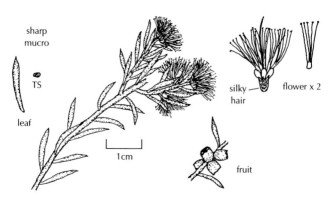

An attractive pink-flowering garden shrub.

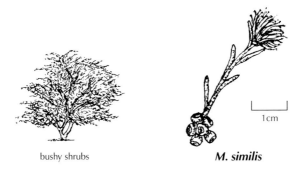

bushy shrubs

M. similis

M. striata Labill.

DESCRIPTION ➤ Normally low, somewhat spreading, straggly shrub, under 1m, although it may grow erect and taller in certain aspects.
LEAVES spirally arranged, falcate to linear-lanceolate, striate, about 15mm long by 2–3mm wide, pungent-tipped; young growth silky-hairy.
FLOWERS pink to mauve–pink, in cylindrical to ovoid heads 20–40mm long by 20–25cm wide; singly or in clusters; hypanthium white-hirsute. Flowering season: mostly early summer.
FRUITS form an ovoid or oblong spike, usually 20–25mm long by over 10mm across.

DISTRIBUTION ➤ Fairly widely distributed in Eyre and Stirling districts of south-west WA, including Stirling Range and Cape Le Grand National Parks.

DISTINGUISHING FEATURES ➤ Very distinctive pungent, falcate to linear–lanceolate leaves with 3 longitudinal nerves.
   This species is being considered for a generic change, as it does not entirely meet the criteria for *Melaleuca*.

CULTIVATION ➤ Not easy to grow. Appears to require acidic, sandy or gravelly to light clay soils in winter-rainfall areas of 400mm or more. Tends to collapse without warning.

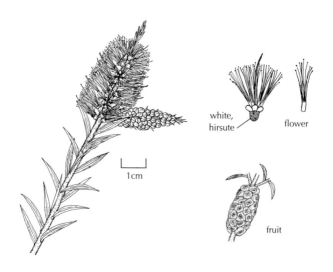

A lovely but tricky shrub for the garden.

low, spreading shrub

M. strobophylla Barlow, in Barlow & Cowley

DESCRIPTION ➤ Mainly a spreading or bushy-crowned shrub or small tree, 4–12m, with grey–white, papery bark.

LEAVES spirally arranged, flat and twisted, narrowly elliptic to sometimes narrowly obovate, 6–15mm long by 1.5–2mm wide, apex sharply acuminate and base contracted into a short petiole; glabrous except for hairy new growth.

FLOWERS creamy white, in quite showy spikes to 60mm long; flowers in dyads on the inflorescence; sepals ovate. Flowering season: January–March.

FRUITS narrowly barrel-shaped, 3–4mm long and wide, with deeply recessed valves, tending to be scattered along branches.

DISTRIBUTION ➤ WA, from Stirling Range area to Salmon Gums district, with a disjunct occurrence in Eneabba–Three Springs and Yalgoo districts, favouring wet, saline depressions in heavy soils.

DISTINGUISHING FEATURES ➤ Mainly small tree proportions with papery bark, sharp-pointed, twisted, mostly narrowly elliptic leaves and showy, creamy white flower spikes in summer, the flowers in dyads on the spike.

Previously included under *M. preissiana* (p. 224) but differs in its twisted, sharp-pointed leaves, fewer stamens per bundle and flowers in dyads with ovate sepals.

CULTIVATION ➤ Found in heavier, saline soils than *M. preissiana* and is a very useful plant in cultivation for these conditions, at least in winter-rainfall semi-dry to temperate areas.

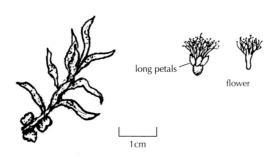

A good-looking paperbark for heavy, slightly saline soils.

small, bushy-crowned tree

M. styphelioides Smith
Prickly Paperbark

DESCRIPTION ➤ Paperbark tree with erect habit and dense, regular canopy of foliage; usually 8–15m high, but reaching 20m at its best; papery, grey–white bark.
LEAVES spirally arranged, ovate to ovate-lanceolate, with a long, pungent apex, usually about 10–25mm long by 3–6mm broad but sometimes larger, glabrescent, with fine, longitudinal nerves.
FLOWERS profuse, cream, in cylindrical spikes about 30mm long by 20mm wide; new leaves growing from end of spike; hypanthia hairy; stamens 12–26 per bundle. Flowering season: November–December.
FRUITS small rounded, woody capsules, about 3mm across.

DISTRIBUTION ➤ A tree of coastal streams and estuaries of southern Qld and NSW, as far south as the Shoalhaven River near Nowra.

DISTINGUISHING FEATURES ➤ Very prickly, ovate to ovate-lanceolate leaves with fine longitudinal nerves.

M. squamophloia (Byrnes) Craven, occurring in black soil plains of the Miles–Jandowae–Tara district of Qld, was previously recorded as a variety of *M. styphelioides*. It is generally a smaller tree distinguished by its quite different leaf oils.

CULTIVATION ➤ *M. styphelioides* seems to grow well almost anywhere, from temperate to sub-tropical areas, in swampy sand or heavy winter wet clay or quite dry soils, where annual rainfall is above about 450mm. It is a successful street tree in Adelaide, and also succeeds in alkaline clay soils. A massive specimen can be seen in the Melbourne Botanic Gardens.

M. squamophloia has rarely been cultivated.

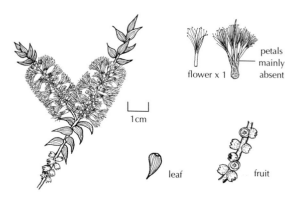

This lovely paperbark succeeds in almost every situation.

erect paperbark tree

M. suberosa (Schauer) C. Gardner
Cork-bark Honey-myrtle

DESCRIPTION ➤ Unusual small shrub, 30cm–1m, rarely spreading, with corky bark and tiny crowded leaves.
LEAVES linear or sub-terete, 2–7mm long, crowded, overlapping or spiralling around branches, tuberculate on upper surface, acute or obtuse.
FLOWERS deep pink, immersed in fissures of the corky bark, occurring in profusion over considerable lengths of branches; with unpleasant odour. Flowering season: late winter or spring.
FRUITS 4–5mm across, square-sided, in rows along the branches.

DISTRIBUTION ➤ South coast of WA, from Albany to Israelite Bay, and inland to near Cocklebiddy.

DISTINGUISHING FEATURES ➤ Small stature, corky bark, tiny leaves and small branchlets, and deep pink flowers clustered along and around the old branches only, over considerable lengths (ramiflorous habit).

SIMILAR SPECIES ➤ *M. agathosmoides* C.A. Gardner occurs in Lake King district of WA. Although unrelated, this dwarf species is included here because it also features a ramiflorous habit, uncommon in the genus. It has tiny decussate leaves 2–3mm long, and white to green–white flowers on lateral inflorescences on the old wood. Flowering season: July–November.

CULTIVATION ➤ Its unusual but beautiful flowering habit makes *M. suberosa* popular amongst native plant enthusiasts, although it is rarely sighted in nurseries. It requires well-drained acidic, sandy or light soils in low humid areas for success and is usually unreliable in other conditions.
M. agathosmoides is unknown to the author in cultivation.

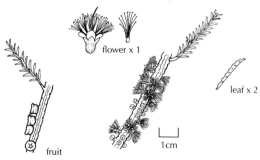
flower x 1
leaf x 2
fruit
1cm

A dwarf shrub that displays a unique, showy flowering habit.

dwarf shrubs

M. agathosmoides

M. subfalcata Turcz.

DESCRIPTION ➤ Spreading or erect, bushy shrub, 0.5–4m, with rough fibrous bark.
LEAVES spirally arranged, glabrous, sub-falcate or linear-recurved, and sharply pointed, normally 8–30mm long by 1–3mm wide.
FLOWERS in dense, oblong spikes about 50mm long by 20–30mm broad; bright pink and profuse at their best, but also pale mauve to nearly white, and not always prolific; hypanthia hairy; stamens 11–22 per bundle. Flowering season: usually October–December.
FRUITS about 5mm diameter, scattered or clustered along branches.

DISTRIBUTION ➤ WA, from Ongerup district eastwards to Israelite Bay.

DISTINGUISHING FEATURES ➤ Crowded, sub-falcate, mostly recurved leaves and pink to mauve lateral flower spikes.

CULTIVATION ➤ Grows well in most soils and situations in semi-dry to temperate areas with rainfall exceeding about 300mm annually. Its response to humidity is unknown to the author. Frost hardy. Usually a low profuse-flowering shrub, it sometimes reaches small-tree proportions with sparser flowers if grown in heavy rich soils of assured rainfall.

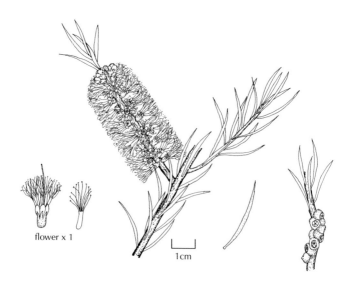

flower x 1

1cm

An easily grown garden shrub with handsome flowers.

dwarf to large shrub

M. systena Craven

DESCRIPTION ➤ Smallish, bushy shrub, 0.2–2m, but rarely exceeding 1m.

LEAVES crowded, often recurved, spirally arranged, narrow-linear, very narrowly obovate or very narrowly elliptic, 4–15mm long by 0.5–1.5mm wide, with a few silky hairs, but often glabrous; where hairs are present they are dimorphic, both branchlets and leaves featuring short hairs overlaid by sparse long ones.

FLOWERS in rounded, dense, terminal and axillary heads 12–20mm wide; profuse; in triads of 3–9; filaments yellow, cream or white, with yellow anthers, overall colour appearing yellow; hypanthia 1.5–2.5mm long, glabrous or hairy, mainly at the base; calyx with ciliate margin. Flowering season: winter–summer, but flowers more likely in spring.

FRUITS mostly urn-shaped capsules, 3–6mm long, in small irregular clusters.

DISTRIBUTION ➤ WA, from Shark Bay district south to Hamelin Bay district, but also further south and inland to Walkaway and Eneabba areas. Occurs in various habitats including coastal scrub, white dune sand and rocky limestone.

DISTINGUISHING FEATURES ➤ Dimorphic hairs on the branchlets and leaves, hypanthia which may be glabrous or hairy, and ciliate margin to the calyx. Leaves and flowers as described.

CULTIVATION ➤ Adaptable species, tolerating most soils including limestone; flowers prolifically in full sun and could be used for near-coastal gardens. Requires good drainage and temperate to semi-dry conditions. Frost hardy. Previously grown as *M. acerosa*, an invalid name.

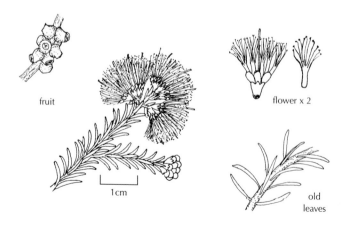

fruit

1cm

flower x 2

old leaves

An adaptable, free-flowering shrub for moist soils, including limestone.

bushy shrub

M. teretifolia Endl.

DESCRIPTION ➤ Erect shrub, normally 3–5 m, with greyish, papery bark.

LEAVES spirally arranged, needle-like, terete, 30–90mm long by 1 mm wide, with straight, acute point.

FLOWERS in globular, sessile, axillary or sometimes lateral clusters of 4–15 monads along branches; usually white, but deep pink to red colour forms known as *M.* 'Georgiana Molloy' are in cultivation (p. 120). Flowering season: late spring and summer.

FRUITS small, rounded capsules 3–4 mm diameter, forming globular clusters along branches.

DISTRIBUTION ➤ Mainly in Darling region of WA, favouring wet, marshy places from Perth to Bunbury, but extending north to Watheroo district.

DISTINGUISHING FEATURES ➤ Thin, needle-like, straight-pointed, smooth leaves, and globular clusters of flowers over long lengths along the branches.

CULTIVATION ➤ Tricky to cultivate away from wet, sandy or gravelly, acidic soils. Resents heavy clay and alkaline soils, where growth is very slow and unhealthy. Unknown by the author on the more humid east coast.

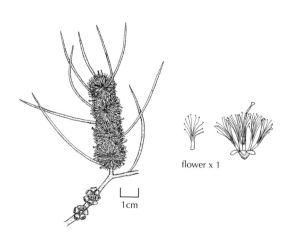

flower x 1

1cm

In suitable soils this can be a spectacular flowering shrub.

tall, erect shrub

M. thapsina Craven

DESCRIPTION ➤ Small to large, prickly shrub, 0.7–4m tall by about 1m wide.
LEAVES spirally arranged, 10–50mm long by 1–1.5mm wide, terete, glabrescent with a very sharply pointed, straight tip.
FLOWERS bright yellow or cream, in tightly packed, rounded or cylindrical heads 20mm or more long by 15–18mm wide; hypanthia hairy; petals deciduous; stamens 6–8 per bundle; 5 distinct calyx lobes. Flowering season: usually September–October.
FRUITS tightly packed into globose or oblong clusters, the calyx lobes weathering away.

DISTRIBUTION ➤ Lake King–Norseman–Ravensthorpe–Esperance district of WA, in varying habitats in clay or sandy soils.

DISTINGUISHING FEATURES ➤ Pungent, needle-like leaves with straight tips; masses of usually bright yellow inflorescences of 2–13 triads, with 5 distinct calyx lobes.

SIMILAR SPECIES ➤ *M. halophila* Craven, from Fitzgerald Peaks–Salmon Gums district of WA, differs in its flatter, hairy and generally shorter leaves, 11–30mm long by 17–25mm wide, scarious calyx lobes and white flowers in spring.

    M. johnsonii Craven, from Hyden–Marvel Loch–Norseman district south to Newdegate–Esperance district of WA, has narrow, flattened, semi-elliptic to quadrate leaves only 7–17mm long, scarious calyx lobes and cream to yellow flowerheads, 13mm wide, in spring.

CULTIVATION ➤ *M. thapsina* is unknown to the author in cultivation but it would make a showy flowering, but very prickly shrub of distinction, if used judiciously. Should adapt to a range of soil types. The other two species are also unknown in cultivation.

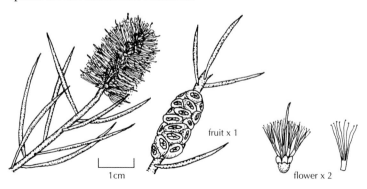

A prickly shrub with spectacular flowers.

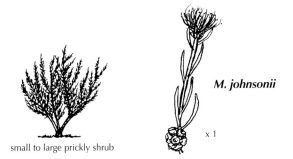

small to large prickly shrub

M. johnsonii

x 1

M. thymifolia Smith

DESCRIPTION ➤ Low, spreading, glabrous shrub, usually under 1m, with rather lax, arching branches.

LEAVES elliptic to narrowly elliptic, acute, 5–15mm long by 1–3 mm wide, mainly in opposite pairs (decussate), often lying flat, or parallel to branches.

FLOWERS in showy clusters at base of secondary branches; petals long and deep violet (also a white flowering form); stamens pinkish mauve to violet, with numerous brush-like filaments on inside of staminal claw. Flowering season: long period, mainly summer.

FRUITS cup-shaped, 3–5mm long and wide, with prominent, persistent, sepaline teeth.

DISTRIBUTION ➤ Coastal areas and Darling Downs of southern Qld, to NSW as far south as Ulladulla and west to the Blue Mountains, and inland to Mudgee area. Also an outlier in Carnarvon Range district of Qld.

DISTINGUISHING FEATURES ➤ Violet to pinkish mauve flower clusters, produced mainly during summer, and staminal bundles, brush-like due to the inward-pointing filaments.

CULTIVATION ➤ One of the most commonly cultivated melaleucas due to its small size, ornamental flowers over a long period, and adaptability to temperate and sub-tropical climates. It requires ample water and resents limy soils. A number of registered cultivars such as 'Cotton Candy', 'White Lace' and 'Pink Lace' are available.

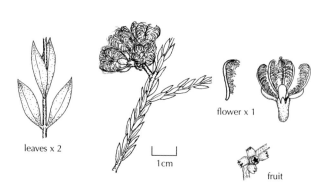

leaves x 2   1cm   flower x 1   fruit

This popular garden shrub is showy and long-flowering.

low, twiggy shrub

M. thymoides Labill.

DESCRIPTION ➤ Low, branching shrub, normally about 1m high and wide, sometimes much taller.

LEAVES spirally arranged, glabrous, narrowly elliptic or lanceolate, 5–7mm long by about 2mm wide, and 3-nerved; smaller branchlets usually end in a sharp spine.

FLOWERS profuse, bright daffodil yellow; in ovoid or globular terminal or lateral heads about 15mm across. Flowering season: October–November.

FRUITS quite large, usually fused into a few-fruited cluster 10mm or more wide.

DISTRIBUTION ➤ Widespread in WA, from Perth to Eyre district, often exposed to sea winds on the foreshore or on coastal cliffs.

DISTINGUISHING FEATURES ➤ Usually spinescent ends to the branchlets, small, pointed, elliptic to lanceolate leaves, bright yellow flowerheads and distinctive fruits.

This species does not entirely meet the criteria for *Melaleuca* and may be separated into another genus.

CULTIVATION ➤ Attractive in flower, but has rarely been cultivated. Attempts indicate this species is difficult to establish, further evidence that it may not belong in *Melaleuca*.

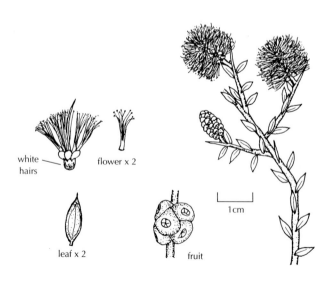

A wiry shrub, with notable yellow flowerheads.

low, branching shrub

M. thyoides Turcz.

DESCRIPTION ➤ Erect, woody shrub, normally 2–5m high and wide, the thinner branches often arching, with rough, dark grey bark.

LEAVES tiny, scale-like, spirally arranged and overlapping, except on older, thicker branches; narrowly ovate on older branches, more or less rhomboid or ovate on thin branchlets.

FLOWERS white with a pink blush, or cream; in small, oblong to ovoid–globular spikes 15–40mm long, the axis growing on; stamens 4–10 per bundle. Flowering season: spring or summer.

FRUITS cup-shaped to globular, 4–6mm diameter, in thick clusters.

DISTRIBUTION ➤ Widely distributed in south-west of WA, from Lake Monger north of Perth, south to Ongerup district and eastwards to Cape Arid.

DISTINGUISHING FEATURES ➤ Tiny, scale-like, spirally arranged leaves on very thin branches and rough, dark grey bark.

SIMILAR SPECIES ➤ ***M. tamariscina*** Hook. (p. 150) from Queensland is very similar, although its thin, young branchlets are very tamarisk-like, with numerous tiny side shoots. Leaf arrangement and fruits are almost identical. It can be separated, however, by its white or creamy brown, papery bark and very distinctive weeping habit.

CULTIVATION ➤ *M. thyoides* is a very useful, tall shrub for swampy, saline or alkaline soils. Although not especially ornamental, the very thin branchlets and profuse flowering have their own appeal. Forms a good windbreak, and is suited to full sun or semi-shade. Frost tolerant. Tolerance to the more humid east coast is unknown.

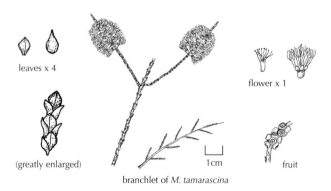

leaves x 4

(greatly enlarged)

branchlet of *M. tamarascina*

flower x 1

fruit

A good shrub to combat saline and alkaline soils.

erect, woody shrub

M. tinkeri Craven

DESCRIPTION ➤ Small, spreading shrub, seldom reaching 1m.
LEAVES linear or linear-obovate, spirally arranged, 8–34mm long by 0.5–1.5mm wide, silky-hairy, warty, the new growth soft and greyish; distinct oil glands are scattered.
FLOWERS pink to magenta; in globular heads to 17mm wide, comprising 4–12 triads; hypanthia and lobes hairy; stamens 3–6 per bundle; style under 10mm long. Flowering season: July–October.
FRUITS small, in tight balls or pineapple-like clusters, lobes usually weathering away; ovules 10–12 per locule.

DISTRIBUTION ➤ Yandanooka–Gairdner Range district of WA.

DISTINGUISHING FEATURES ➤ Very warty, hairy leaves, coupled with relatively small, pinkish inflorescences and compacted, globose fruiting clusters.

SIMILAR SPECIES ➤ *M. clavifolia* Craven, from Coorow–Green Head district of WA south to Moore River district, is a similar small shrub, differing by its generally shorter leaves (4–10mm), with a dimorphic hair arrangement, wider inflorescences (to 23mm), more ovules per locule (15–20) and less distinct oil glands in rows. The leaf apex is generally more rounded, giving a club-shaped appearance.

CULTIVATION ➤ The author has no knowledge of either species being cultivated but they should perform in similar manner to other species from the general area (e.g. *M. trichophylla*, p. 296).

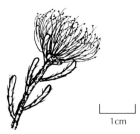

M. clavata

A free-flowering low bushy, or prostrate, shrub that makes a good groundcover.

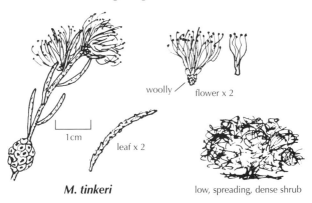

M. tinkeri

woolly — flower x 2

leaf x 2

low, spreading, dense shrub

M. torquata Barlow

DESCRIPTION ➤ Medium-sized, bushy, very prickly shrub, to 2.5m, with rough bark.

LEAVES very prickly, more or less glabrous, mucro sharp; spirally arranged; 5–12mm long by 1–2mm wide, narrowly elliptic to narrowly ovate, often reflexed, with prominent keel on undersurface; young growth very hairy.

FLOWERS profuse, white, petals tinged pink, in rounded heads 15–18mm across, mainly terminal on short side branchlets; buds reddish pink; hypanthia glabrous; stamens 3–13 per bundle. Flowering season: September–October

FRUITS about 5mm long and wide, more or less cup-shaped, sepals usually persistent as sepaline teeth; in small clusters or singly.

DISTRIBUTION ➤ Southern WA, mainly in heavy soils in eucalypt associations, extending from Tambellup to just east of Esperance (Mt Ney).

DISTINGUISHING FEATURES ➤ Very prickly, flat, keeled leaves, often reflexed, and globular, terminal, white-tinged pink flowerheads, usually less than 10 stamens per bundle.

SIMILAR SPECIES ➤ *M. teuthidioides* occurs from Marvel Loch district of WA south and eastwards to Ravensthorpe and Balladonia, mainly as eucalypt understorey. It is a similar rough-barked shrub, differing in its non-keeled, shorter leaves, hairy hypanthia, more stamens per bundle (12–16), and calyx lobes on the fruits which weather away.

M. linguiformis Craven, from Salmon Gums–Wittenoom district of WA, is close to *M. teuthidioides*, differing by its shorter calyx lobes and, generally, more stamens per bundle (13–22).

CULTIVATION ➤ All three species grow on a range of soils and should adapt to light or fairly heavy soils, acid or alkaline, in semi-dry to temperate conditions. *M. torquata* is growing successfully in alkaline loam in Adelaide's northern district. The author knows of no other examples in cultivation.

bushy, prickly shrub

This shrub is showy in flower and good for harsh conditions.

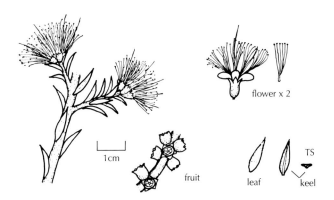

M. trichophylla Lindl.

DESCRIPTION ➤ Low, ground-hugging, spreading shrub to 50cm, or more erect, bushy shrub to 1m or more.

LEAVES narrow, linear to linear-obovate, 8–31mm long by about 1mm wide, spirally arranged, with prominent, usually scattered, oil glands; sometimes glabrous, but more often feature dimorphic hairs, with shorter hairs overlaid by sparser, longer, pubescent hairs.

FLOWERS prolific, varying from pink to purple or rich carmine, tipped by golden anthers; in clustered terminal and upper axillary heads to 35mm wide; flowers comprise 2–12 triads; hypanthia and lobes hairy; stamens 5–11 per bundle; styles long (10–20mm); papery brown bracts overlap the unopened buds. Flowering season: August–December.

FRUITS in non-globose small clusters, often subtended by persistent leaf bases. Capsules 2–5mm long.

DISTRIBUTION ➤ Very variable WA species, extending from Northampton district south to Busselton district.

DISTINGUISHING FEATURES ➤
Profuse, very showy inflorescences subtended by papery brown bracts, and warty, hairy, sometimes dimorphic, narrow leaves, coupled with non-globose fruiting clusters.

SIMILAR SPECIES ➤ *M. beardii* Craven, from Ajana–Carnamah region of WA, is a related small shrub featuring generally shorter, dimorphic leaves with an obtuse apex and similar, smaller flowerheads of 3–6 triads.

CULTIVATION ➤ *M. trichophylla* has long been cultivated, one form which flowers in August being particularly adaptable to a range of soils. The showy, deep magenta form which flowers in late November–December tends to be short-lived, a pity as it is particularly attractive in full flower. Well-drained acidic to slightly alkaline soils in warm temperate regions appear best.

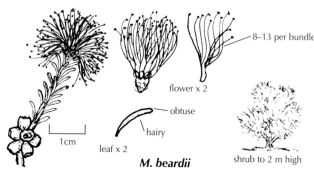

8–13 per bundle

flower x 2

obtuse

hairy

leaf x 2

M. beardii

1cm

shrub to 2 m high

Upright form of *M. trichophylla*

M. beardii

A groundcover or bushy shrub, spectacular in flower.

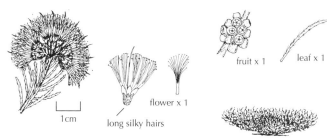

M. trichophylla

long silky hairs

flower x 1

fruit x 1

leaf x 1

ground-hugging shrub

M. tuberculata Schauer, in Lehm.

DESCRIPTION ➤ All 3 varieties of *M. tuberculata* are normally small shrubs, seldom larger than 90cm high and wide, but occasionally larger.
LEAVES very variable in size, resulting in division into three varieties; spirally arranged, mostly blue–green to greyish, narrowly obovate, obovate, elliptic or linear, warty on undersurface.
FLOWERS profuse, pale to mid-pink or mauve, in gold-tipped heads 15–25mm across; hypanthia and lobes woolly-hairy; brown buds clothed in white, woolly hairs; hairy brown bracts subtend flowers; petals caducous. Flowering season: normally spring.
FRUITS cylindrical to urceolate, about 4mm long, in small irregular clusters, calyx lobes weathering away.

DISTRIBUTION ➤ Mainly inland WA species (see individual varieties).

DISTINGUISHING FEATURES ➤ Bluish, narrowly obovate but variable leaves, buds clothed in white, woolly hairs, woolly hypanthia and lobes.

Variety *arenaria* (C.A. Gardner) Craven, occurring in Kulin–Hyden–Pingaring district, has small leaves 2–4mm long by 1–2.3mm wide.
  Variety *macrophylla* Craven, from Kulin–Jerramungup district eastwards to Grass Patch–Israelite Bay area; has leaves 4–18mm long by 1–3.5mm wide.
  Variety *tuberculata*, from Brookton–Narrogin district east to Esperance–Israelite Bay district, has leaves 2.5–13.5mm long by 0.8–1.3mm wide.

SIMILAR SPECIES ➤ *M. leptospermoides* Schauer, in Lehm, found from Cadoux–Brookton district eastwards to Coolgardie–Lake King district of WA, is another very small shrub with very similar features to *M. tuberculata* but lacking the hairy brown bracts, and having slightly different hairs on the calyx lobes.

CULTIVATION ➤ The author has no evidence of any of these shrubs being cultivated but their habitat range suggests they should grow well in most acid–mildly alkaline soils in temperate areas. Good forms of *M. tuberculata* are particularly showy in full flower.

dwarf shrubs

M. leptospermoides

M. tuberculata var. *macrophylla*

Excellent potential as attractive garden shrubs.

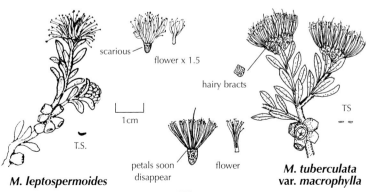

M. uncinata R. Br., in Ait.
Broombush

DESCRIPTION ➤ Broom-like shrub, normally 1–4m, with numerous ascending branches and brownish, papery to smooth bark.
LEAVES spirally arranged, terete or linear-terete, glabrescent except on silky new shoots, 10–30 long by 1–4 wide, tapering to distinct curved or hooked, pointed apex.
FLOWERS cream to yellow, in dense, axillary, globular or oblong heads, usually 15–17mm across; hypanthia silky-hairy. Flowering season: mainly spring.
FRUITS smooth capsules, usually under 3mm diameter, in tightly compacted clusters.

DISTRIBUTION ➤ Widespread, occurring in all mainland States mainly south of Tropic of Capricorn. Well known for its use in brush fencing and shadehouses.

DISTINGUISHING FEATURES ➤ Brush-like or broom-like habit with papery bark and terete-type leaves tapering to a distinct hooked point (uncinate).

SIMILAR SPECIES ➤ *M. stereophloia* Craven, from Wooramel–Meekatharra districts south to the Coorow and Koorda districts in WA, is closely related, differing in its hard, grey, fibrous bark, amongst other features.

The *M. uncinata* complex is being studied in more detail and new species names are likely to arise from this study.

CULTIVATION ➤ *M. uncinata* can be grown in most well-drained soils and situations in dry to temperate climates. Normally it is not particularly ornamental, although good, profuse, flowering forms are sometimes encountered over its wide range. Useful for hedging or windbreaks. Frost hardy.

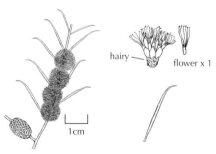

A useful shrub for moderately dry conditions.

broom-like shrub

M. urceolaris F. Muell. ex Benth.

DESCRIPTION ➤ Low, soft-foliaged shrub, usually no more than 1m high, spreading to 1m or more by arching branches low to the ground.
LEAVES crowded, thin, spirally arranged, mostly linear, green, but of greyish appearance due to covering of softly pubescent hairs; mainly 7–20mm long by 0.5–1mm wide, incurved, sometimes spirally twisted; leaves on branchlets are in 2 layers, the longer, outer layer quite sparse.
FLOWERS profuse, creamy white or yellow, turning pinkish red as they age, in heads of 2–12 monads, to 25mm across. Flowering season: normally September–November, sometimes as early as July.
FRUITS in non-globose, peg-like clusters.

DISTRIBUTION ➤ Arrino–Jurien Bay–Gingin districts of WA.

DISTINGUISHING FEATURES ➤ Crowded, mainly incurved leaves clothed in soft pubescent hairs, and creamy to yellow flowerheads ageing to red, comprising 2–12 monads.

SIMILAR SPECIES ➤ *M. virgata* (Benth.) Craven, overlapping in habitat in Arrino–Hill River–Moora districts of WA, is somewhat similar. Its cream to yellow inflorescences age to pinkish red, but differ in comprising 1–4 triads. Leaves are also slightly shorter (4–16mm long) and greener.

CULTIVATION ➤ *M. urceolaris* is a handsome foliage shrub, spectacular in flower in good forms, which grows successfully in winter-rainfall temperate to semi-dry regions in deep sand and clay.

M. virgata would probably respond similarly but is unknown to the author in cultivation.

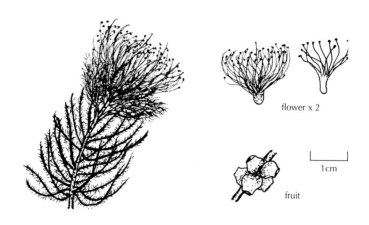

flower x 2

fruit

1cm

M. virgata

Spectacular in flower and with lovely soft greyish foliage.

x 1   ***M. virgata***   low, spreading shrub

M. venusta Craven

DESCRIPTION ➤ Branching, open shrub, 0.5–1.5m.
LEAVES silky-grey to grey–green, oblong, obovate or elliptic, flat, with a small mucro, 12–45mm long by 5–10mm wide, spirally arranged and petiolate; dense silky hairs cover both surfaces; veins parallel.
FLOWERS pinkish mauve and fading rapidly to white, in moderately large heads to 32mm wide, comprised of 6–13 triads; hypanthia silky. Flowering season: Late spring–early summer.
FRUITS form a globular or pineapple-shaped tight cluster.

DISTRIBUTION ➤ Known only from a small coastal area north of the Murchison River in WA, in sand over limestone.

DISTINGUISHING FEATURES ➤ Silky, greyish leaves similar to *M. conothamnoides* (p. 64) except for colour and texture.

SIMILAR SPECIES ➤ *M. keigheryi* Craven, from further north in Shark Bay district of WA, is a somewhat similar shrub to 2m or more tall, with many similar features to *M. venusta*. The leaves, however, are much greener, to 23mm long, while the pink, fading white flowerheads to 25mm wide are comprised of 4–9 triads. The leaf blade differs distinctly by its pinnate veins.

CULTIVATION ➤ Although *M. venusta* has proven adaptable to various soil types, at least in Adelaide, the rapid deterioration of its flowers, and very loose, open habit make it disappointing as a garden plant.
  The author has no knowledge of *M. keigheryi* in cultivation.

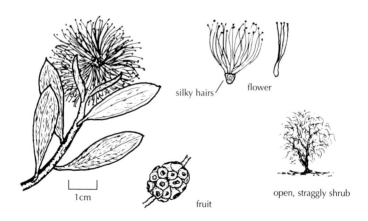

deep colour form

Lovely foliage and flowers but a very open straggly habit.

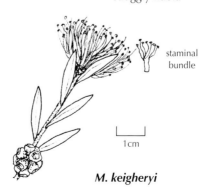

staminal bundle

M. keigheryi

M. viminea Lindl. subsp. *viminea*

DESCRIPTION ➤ Small to large, densely branched, variable shrub or small tree, to 14m tall by 2–4m wide, with fibrous or papery bark.
LEAVES alternate to ternate, spreading, glabrescent, 5–20mm long by 1–2mm wide, of varying near-linear shapes, untwisted, recurved to straight, the apex pointed.
FLOWERS profuse, white to cream and sickly smelling, in spikes or heads 20–40mm long by 20–25mm wide; hypanthia mainly glabrous; stamens up to 15 per bundle. Flowering season: may occur July to November.
FRUITS in spike of cupular capsules, each capsule 3–5mm wide with sepaline teeth, or the lobes weathering away.

DISTRIBUTION ➤ Widespread in south-west WA in variable forms, from Kalbarri south to Busselton and Albany districts. Naturalised locally in southern Victoria.

DISTINGUISHING FEATURES ➤ Sickly smelling, profuse white to cream inflorescences and mostly recurved, narrow leaves.

M. viminea subsp. *appressa* Barlow, in Quinn, Cowley, Barlow & Thiele, occurring in 3 disjunct populations in Ongerup, Mt Burdett and Yilgarn districts of WA, is easily separated by its small elliptic to obovate, appressed leaves.

M. viminea subsp. *demissa* Quinn ex Craven, mainly found in Walpole–Manypeaks district of WA, has leaves similar to subsp. *viminea*, but shorter and strongly recurved.

CULTIVATION ➤ Subsp. *viminea* and subsp. *demissa* are adaptable plants, succeeding under most temperate conditions but requiring an annual rainfall of at least 400mm, preferably more. The rarer subsp. *appressa* is unknown to the author in cultivation.

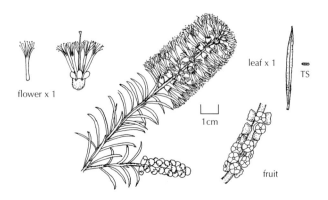

flower x 1   leaf x 1   TS   1cm   fruit

Profuse flowering shrubs for most temperate conditions.

subsp. *demissa*

1cm

subsp. *appressa*

large, dense shrub

M. violacea Schauer, in Lehm.

DESCRIPTION ➤ Wide-spreading, prostrate to semi-prostrate shrub, with many horizontal branches in layers.

LEAVES opposite, glabrous, sessile or slightly petiolate, cordate-ovate to oblong but variable, normally 5–15mm long.

FLOWERS occur laterally on old wood, and in axillary clusters along branches; purple or violet, with smooth calyx tubes and lobes; petals reddish or pink. Flowering season: usually spring.

FRUITS sessile, 4–6mm diameter, star-like in end-view due to large, persistent sepals.

DISTRIBUTION ➤ Southern WA, from Ravensthorpe westwards to Walpole and inland to the Stirling Range and Ongerup district.

DISTINGUISHING FEATURES ➤ Horizontal branching habit, normally ovate-cordate to oblong leaves in opposite pairs, and purplish flowers clustered along the branches.

CULTIVATION ➤ Although not often cultivated, *M. violacea* forms an interesting, spreading, prostrate groundcover with attractive flowers. Grows well in a range of well-drained soils in temperate areas where rainfall exceeds about 400mm annually. May struggle on limestone.

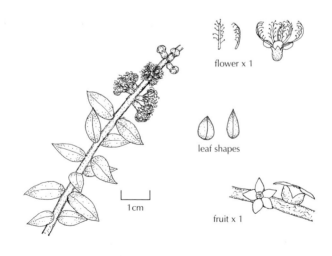

A spreading, prostrate shrub with attractive flowers.

flat, layered shrub

M. viridiflora Sol. ex Gaertn.
Broad-leaved Paperbark

DESCRIPTION ➤ Usually a small, erect or straggly tree, to 10m, with an open canopy and brownish white or grey, papery bark.

LEAVES elliptic, ovate or obovate, thick and leathery, spirally arranged, 60–195mm long by 20–75mm wide, with 5–9 longitudinal conspicuous veins, the apex acute or obtuse; silky hairs on young leaves appressed.

FLOWERS in loose, cylindrical spikes to 100mm long by 55mm wide; spike formed from 8–25 triads; red, or greenish tones of white, cream or yellow, or plain green, white, cream or yellow; rachis, hypanthia and calyx lobes usually hairy; stamens 5–9 per bundle. Flowering season: may be at any time, more often in winter.

FRUITS cylindrical to cup-shaped, 5–6mm in diameter, loosely arranged along branches; seed shed annually.

DISTRIBUTION ➤ Widespread around coastal regions of northern Australia, from Maryborough district in Qld through Arnhem Land to the Dampier Peninsula in WA. Occurs in Kakadu National Park. Often found in summer-wet habitats.

DISTINGUISHING FEATURES ➤ Large, usually greenish yellow to red flower spikes, thick, wide, prominently veined large leaves, young leaves clothed in appressed, silky hairs. Has affinity to *M. quinquenervia* (p. 238) but differs distinctly by the appressed hairs on the young shoots.

CULTIVATION ➤ Good paperbark for the tropics and sub-tropics where it will grow in most soils and habitats, especially summer-wet ones. In nature, its open canopy allows many epiphytes such as orchids to settle, and this feature can be encouraged in gardens.

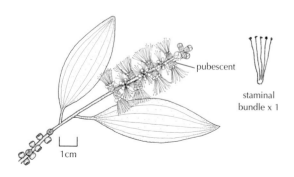

pubescent

staminal bundle x 1

1cm

A good tree for the tropics, with showy flowers.

red-flowering form

small tree

M. wilsonii F. Muell.
Wilson's Honey-myrtle

DESCRIPTION ➤ Dense, widely spreading shrub, 1–2m high by sometimes 3m or more in spread, the older branches becoming leafless and corky. A dwarf form is also in cultivation.

LEAVES decussate, sessile, linear-lanceolate or narrowly ovate, 8–15mm long by 1–2mm broad.

FLOWERS deep pink, in lateral clusters along previous year's branches over considerable lengths, making an attractive display. Flowering season: spring.

FRUITS small, 3–4mm across, scaly, with persistent sepals.

DISTRIBUTION ➤ South-east SA, particularly Keith–Coonalpyn area, and north-west Victoria, in sandy soils.

DISTINGUISHING FEATURES ➤ Narrow, sharply pointed, decussate leaves, clusters of deep pink flowers along the branches, usually in September–October, and long rows of fruiting capsules along the old wood.

CULTIVATION ➤ Frequently cultivated, this shrub thrives in light and heavy soils in winter-wet temperate areas. Suited to alkaline soils as well as most others. Frost tolerant.

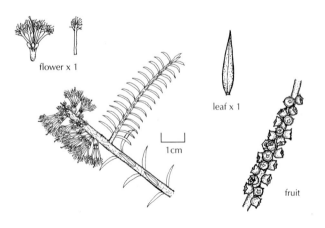

A spectacular shrub when in flower.

wide-spreading shrub

M. xerophila Barlow, in Barlow & Cowley

DESCRIPTION ➤ Large shrub or small spreading tree, 3–6m, with fibrous or papery bark.

LEAVES glabrous except for tomentose young shoots, spirally arranged, narrowly elliptic, terete or sub-terete, with small, reflexed mucro, fleshy; sharply contracted at base into a short petiole.

FLOWERS profuse, white or cream; in small, mainly terminal, leafy heads, sometimes in axils at ends of branchlets, the axis rarely growing on; stamens 15–22 per bundle. Flowering season: September–October.

FRUITS small, 2.5–3.5mm long and wide, cup- or barrel-shaped, singly or in clusters.

DISTRIBUTION ➤ Scattered arid locations in WA and SA, usually in depressions adjoining salt lakes.

DISTINGUISHING FEATURES ➤ Small, fleshy, near-terete, mucronate leaves, and small but numerous white or cream inflorescences on a woolly axis. Closely related to *M. lanceolata* (p. 154).

CULTIVATION ➤ Although virtually unknown in cultivation, a healthy specimen growing at the Waite Arboretum in Adelaide on calcareous clay soil, together with its natural habitat, suggests *M. xerophila* may be a good plant for difficult limestone soils, at least in winter-rainfall areas.

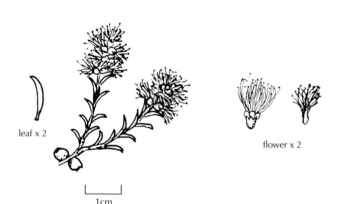

leaf x 2

flower x 2

1cm

A good groundcover shrub with brilliant flowers.

A useful tree for arid gardens and saline depressions.

tall shrub or small tree

FLOWERS probably pale to mid pink with large golden anthers, in quite large terminal heads to 35mm across; flowerheads subtended by large, ovate, concave, brown bracts persisting for some time after flowers open; hypanthia softly hairy. Flowering season: usually late October–early November.

FRUITS cup-shaped, about 4mm long by 4–5mm wide, mostly in small peg-like clusters.

DISTRIBUTION ➤ Unknown, probably of garden origin.

SIMILAR SPECIES ➤ Prominently ciliate and striate linear (younger) to oblanceolate (older) leaves and large, showy, pale pink, yellow-tipped, terminal flowerheads. Appears to be related to *M. psammophila* (p. 228).

CULTIVATION ➤ Dense, small, foreground shrub which grows well in Adelaide in sand or clay of acidic or slightly alkaline reaction. Flowers are very attractive, but last only for 2–3 weeks. The author has no knowledge of it being grown elsewhere.

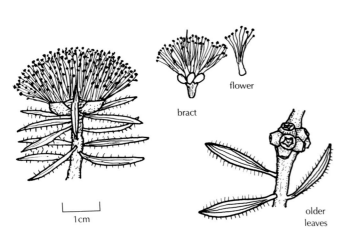

M. un-named species 2

DESCRIPTION ➤ Erect, slender shrub, to 1.5m, with many ascending branches.
LEAVES spirally arranged and widely spaced, mainly 30–40mm long by about 3mm wide, glabrous, slightly pungent, linear to linear-oblanceolate, flat with prominent margins and midrib.
FLOWERS blush pink to salmon pink, in globular, terminal heads about 16–20mm in diameter. Flowering season: October.
FRUITS small capsules, tightly packed in oblong heads.

DISTRIBUTION ➤ Specimen described was found by the author at Wittenoom Hills north of Esperance, WA; any wider distribution unknown.

DISTINGUISHING FEATURES ➤ Long, smooth, flat, linear to linear-oblanceolate leaves and globular, blush pink to salmon flowerheads. The northern WA species *M. concreta* (p. 62) has similar leaves but differs in its smaller, yellow, mostly lateral flowerheads.

CULTIVATION ➤ Has never been found since a bushfire occurred in the area some time after its discovery, and has never been cultivated.

A useful tree for arid gardens and saline depressions.

tall shrub or small tree

M. un-named species 1

DESCRIPTION ➤ Low, spreading shrub, seldom more than 90cm high by 1m wide.

LEAVES linear to oblanceolate, pustulate, spirally arranged, to 25mm long by 4mm wide, striate, prominently ciliate, spreading and often reflexed.

FLOWERS profuse, pale to mid-pink with large golden anthers, in quite large terminal heads to 35mm across; flowerheads subtended by large, ovate, concave, brown bracts persisting for some time after flowers open; hypanthia softly hairy. Flowering season: usually late October–early November.

FRUITS cup-shaped, about 4mm long by 4–5mm wide, mostly in small peg-like clusters.

DISTRIBUTION ➤ Unknown, probably of garden origin.

SIMILAR SPECIES ➤ Prominently ciliate and striate linear (younger) to oblanceolate (older) leaves and large, showy, pale pink, yellow-tipped, terminal flowerheads. Appears to be related to *M. psammophila* (p. 228).

CULTIVATION ➤ Dense, small, foreground shrub which grows well in Adelaide in sand or clay of acidic or slightly alkaline reaction. Flowers are very attractive, but last only for 2–3 weeks. The author has no knowledge of it being grown elsewhere.

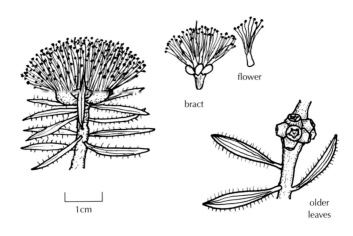

A good groundcover shrub with brilliant flowers.

low shrub

M. un-named species 2

DESCRIPTION ➤ Erect, slender shrub, to 1.5m, with many ascending branches.
LEAVES spirally arranged and widely spaced, mainly 30–40mm long by about 3mm wide, glabrous, slightly pungent, linear to linear-oblanceolate, flat with prominent margins and midrib.
FLOWERS blush pink to salmon pink, in globular, terminal heads about 16–20mm in diameter. Flowering season: October.
FRUITS small capsules, tightly packed in oblong heads.

DISTRIBUTION ➤ Specimen described was found by the author at Wittenoom Hills north of Esperance, WA; any wider distribution unknown.

DISTINGUISHING FEATURES ➤ Long, smooth, flat, linear to linear-oblanceolate leaves and globular, blush pink to salmon flowerheads. The northern WA species *M. concreta* (p. 62) has similar leaves but differs in its smaller, yellow, mostly lateral flowerheads.

CULTIVATION ➤ Has never been found since a bushfire occurred in the area some time after its discovery, and has never been cultivated.

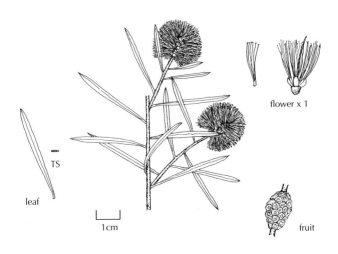

This melaleuca features flowers with a rare colouring.

erect, slender shrub

Glossary

acuminate tapering to a point—referring to the tip of a leaf, the apex angle is sharper or more acute than for an acute tip.

acute sharply pointed, at an acute angle.

adnate joined with another part over the part's whole length.

addressed, appressed pressed close to, or lying flat against something—e.g. referring to the leaves on stems.

alternate placed at different levels—e.g. referring to the position of leaves on branchlets.

anther the pollen-bearing part of a stamen.

apex the tip of an organ.

apiculate with a short pointed tip.

appendage an attached secondary part.

axil the angle between a part and its parent body—e.g. between a leaf and the main stem of a plant.

axillary borne in the axil.

bract a leaf-like structure surrounding a flower or flowerhead.

bracteole a small bract immediately below the calyx of a flower.

caducous falling off early.

calothamnus-like resembling the genus *Calothamnus*, another genus of Australian plants.

calyx (pl. calyces) collective name for the sepals, the whorl of floral parts below the petals.

capitate enlarged and head-like.

capsule a dry fruit formed from a multi-carpelled ovary.

carpel the female part of a flower which bears the ovules; three carpels make up the ovary in *Melaleuca*.

chlorotic yellowing (of the leaves) due to mineral deficiencies, often caused by calcareous soils.

ciliate fringed with hairs.

clavate club-shaped.

conspecific of the same species.

cordate heart-shaped, especially referring to the base of an organ.

corolla collective name for the petals of a single flower.

cuneate wedge-shaped.

cupular cup-shaped.

cylindrical shaped like a cylinder with sides more or less parallel.

decussate arranged in opposite pairs, each pair at right angles to its following pair—referring to the leaf arrangements on a stem.

dimorphic in two different forms.

divaricate branching widely, spreading in different directions in a net-like manner.

dyad pair.

elliptic shaped like an ellipse—e.g. referring to leaves.
elliptic-oblong shaped roughly halfway between elliptic and oblong.
endemic restricted to a particular country or region.
falcate sickle-shaped; flat, curving and tapering to a point.
filament the stalk of an anther.
glabrescent becoming glabrous gradually.
glabrous smooth, without hairs.
glaucous bluish green, with a powdery bloom.
globular rounded or ball-shaped.
hirsute covered with long stiff hairs.
hoary whitish grey; covered with short, dense, greyish hairs.
hyaline delicately membranous and transparent.
hypanthium (pl. **hypanthia**) the floral cup or tube, the base of the flower from which the sepals, petals and stamens arise.
imbricate overlapping.
incurved turned or curved inwards.
inflorescence the flower-bearing system.
keeled a ridge resembling the keel of a boat.
lanceolate lance-shaped; elongated, tapering at each end and broadest below the middle.
lateral occurring at the side.
lateritic of soil, containing laterite or ironstone.
linear long and narrow with more or less parallel sides.
margin the edge (of a leaf).
midrib the main central vein of a leaf, especially when raised.
monad a 'group' of one.
mucro (adj. **mucronate**) a sharp terminal point.
nerve vein.
oblanceolate inversely lanceolate or lance-shaped, the broadest section being above the middle.
oblong with sides more or less parallel except at the base and apex.
obovate inversely ovate; egg-shaped, with the narrow end at the bottom—e.g. referring to the shape of a leaf.
obtuse rounded-ended or blunt.
oil glands organs secreting oil in leaves.
orbicular circular or disc-shaped.
ovary the part of the flower containing the ovules which, when fertilised, becomes the fruit.
ovate egg-shaped, tapering at each end and broadest below the middle—e.g. referring to the shape of a leaf.
ovoid egg-shaped, the broadest part below the middle—e.g. referring to a solid part of a plant, such as the fruit.
ovule the site of egg-cell formation in a plant; the young seed in the ovary prior to fertilisation.
pedicel the stalk of an individual flower.

peduncle the stalk of a cluster of flowers, or of an individual flower if this is the only member of the inflorescence.

peltate of a leaf (or other flat organ), having the stalk attached to its undersurface instead of the edge.

petiolate supported on a petiole.

petiole the stalk of a leaf.

pinnate with the parts arranged on both sides of a central stalk in feather-like fashion.

pseudoterminal terminal inflorescences with new leaves growing on; applies to most melaleucas.

pubescent covered with short, soft or silky hairs.

punctate-glandular dotted with translucent dots or glands.

pungent terminating in a stiff, sharp point.

pusticulate covered with tiny blisters.

pustular blister-like or bearing blisters.

rachis the primary axis of an inflorescence or a compound leaf.

ramiflorous (habit) cluster of flowers along and around the old branches over considerable lengths.

recurved curved backwards.

reflexed bent sharply backwards.

rhomboid, rhombic rhomboid-shaped, resembling an equilateral parallelogram with acute angles; diamond-shaped.

scabrous rough to touch.

scarious thin and dry, and usually not green.

sclerophyll forest dominated by evergreen sclerophyllous trees (trees with hard-textured leaves such as *Eucalyptus*).

sepal one of the parts of the calyx; usually green, protects the softer parts of the flower bud.

sepaline teeth unformed woody protuberances on the fruit.

sessile without a stalk.

spathulate spoon-shaped; more or less oblong or elliptic at the apex, but narrowing significantly towards the base—referring to the shape of the leaves.

spike an arrangement of unstalked flowers attached directly to a common axis.

spinescent ending in a long sharp point.

stamen the male part of a flower, consisting of a filament (stalk) and a pollen-bearing anther.

staminal claw the base of the staminal bundle where the free stamens are joined.

stigma the part of a carpel which receives the pollen.

striate having fine longitudinal lines (striae)—e.g. of leaves with longitudinal veins.

style the sterile part of a carpel connecting a stigma to its ovary.

subfalcate nearly falcate.

subulate awl-shaped.
terete needle-like.
ternate arranged in threes.
tomentose densely covered with short, soft, matted hairs.
tomentum a mat or covering of dense, woolly hairs.
triad in threes.
truncate cut off squarely and abruptly, as in the apex of some leaves.
TS transverse section, cross-section.
tuberculate (n. **tubercle**) having small, wart-like protuberances.
umbrageous providing shade, umbrella-like.
uncinate hooked.
understorey the lower, secondary vegetation to tree species in forest or woodland.
undulate wavy (up and down in different planes).
urceolate urn-shaped.
valve the segment of a fruit which naturally opens at maturity, usually containing seeds.
vein a strand of conducting tissue in the structure of a leaf.
venation the way in which veins are arranged.
verrucose rough and warty.
versatile attached about the middle and free to turn, referring to some anthers, including those of *Melaleuca*.
villous covered with long, soft hairs.
whorled verticillate; arranged in radial formation around the axis.

Bibliography

Australian Plants, Journal of the Society for Growing Australian Plants, various issues.

Barlow, B.A. (1987). 'Contributions to a revision of *Melaleuca*, 1–3'. *Brunonia* 9(2): 163–77.

Barlow, B.A. and Cowley, K.J. (1988). 'Contributions to a revision of *Melaleuca* (Myrtaceae), 4–6'. *Aust. Syst. Bot.* 1: 95–126.

Beard, J.S. (ed.) (n.d.). *West Australian Plants*. Society for Growing Australian Plants, Sydney.

Blackall, W.E. and Grieve, B.J. (1980). *How to Know Western Australian Wildflowers*, part IIIA, 2nd edn, restructured and revised by B.J. Grieve. University of Western Australia Press, Perth.

Brock, J. (2001). *Native Plants of Northern Australia*, revised edn., New Holland, Sydney.

Byrnes, N.B. (1985). 'A revision of *Melaleuca* L. (Myrtaceae) in northern and eastern Australia, 2'. *Austrobaileya* 2(2): 131–46.

Clark, M. and Traynor, S. (1987). *Plants of the Tropical Woodland*. Conservation Commission of the Northern Territory, NT Government Printer, Darwin.

Cowley, K.J., Quinn, F.C., Barlow, B.A. and Craven, L.A. (1990). 'Contributions to a revision of *Melaleuca* (Myrtaceae), 7–10'. *Aust. Syst. Bot.* 3: 165–202.

Craven, L.A. and Barlow, B.A. (1997). 'New taxa and new combinations in *Melaleuca* (Myrtaceae)'. *Novon* 7: 113–19.

Craven, L.A. (1989). 'Reinstatement and revision of *Asteromyrtus* (Myrtaceae)'. *Aust. Syst. Bot.* 1: 373–85.

Craven, L.A. (1998). 'A result of the 1996 Mueller Commemorative Expedition to Northwestern Australia: *Melaleuca triumphalis* sp. nov. (Myrtaceae)'. *Muelleria* II: 1–4.

Elliot, R. and Jones D. (1993). *Encyclopaedia of Australian Plants Suitable for Cultivation*, Vol.6, Lothian, Melbourne.

Holliday, I. (1989). *A Field Guide to Melaleucas*, Hamlyn, Sydney— revised to *A Field Guide to Australian Native Flowering Plants: Melaleucas* (1996), Lansdowne, Sydney.

Holliday, I. (1997). *A Field Guide to Melaleucas*, Vol.2, Adelaide.

Holliday, I. and Watton, G. (1983). *A Field Guide to Australian Native Shrubs*, reprinted edn., Rigby, Adelaide.

Holliday, I. (2002). *A Field Guide to Australian Trees*, 3rd edn., Reed New Holland, Sydney.

Jessop, J. (ed.) (1981). *Flora of Central Australia*, Australian Systematic Botany Society and Reed Books, Sydney.

Jessop, J.P. and Toelken, H.R. (eds) (1986). *Flora of South Australia*, 4th edn., Part II. SA Government Printer, Adelaide.

Marchant, N.G. and others (1987). *Flora of the Perth Region*, Part 1. WA Herbarium, Department of Agriculture, Perth.

Newbey, K. (1968, 1972). *West Australian Plants for Horticulture*, Parts I and II. Society for Growing Australian Plants, Sydney.

Quinn, F.C., Cowley, K.J., Barlow, B.A. and Thiele, K.R. (1992). 'New names and combinations for some *Melaleuca* (Myrtaceae) species and subspecies for the southwest of W.A. considered rare or threatened'. *Nuytsia* 8(3): 333–50.

Stanley, T.D. and Ross, E.M. (1986). *Flora of South-Eastern Queensland*, Vol. II, Queensland Department of Primary Industries, Miscellaneous Publication QM84007, Queensland Government, Brisbane.

Williams, K.A.W. (1979, 1984). *Native Plants of Queensland*, Vols I and II. Author, Brisbane.

Wrigley, J.W. and Fagg, M. (1993). *Bottlebrushes, Paperbarks and Tea-trees and all other Plants in the Leptospermum Alliance*, Angus & Robertson, Sydney.

Index of Botanical and Common Names

Bold = main entry

acacioides 14
acerosa 280
acuminata **12**,
    subsp. *acuminata* **12**
    subsp. *websteri* 12,
adenostyla 216
adnata 100
agathosmoides 276
alsophila **14**
alternifolia 7, **16**
amydra 248
apodocephala
    subsp. *apodocephala* **18**
    subsp. *calcicola* 18
apostiba 162
araucarioides **20**
arcana 68
argentea 8, **22**
armillaris
    subsp. *armillaris* **24**, 104, 136
    subsp. *akineta* 24
aspalathoides **26**
barlowii 64, 196
basicephala 12
beardii 296
biconvexa 52
bisulcata **228**
Black Tea-tree **30**, **154**
blaeriifolia 9, 20, **28**, 86
boeophylla 110
borealis 202
Boree 210
Bracelet Honey-myrtle **24**

bracteata 7, **30**
    'Golden Gem' 30
    'Revolution Gold' 30
    'Revolution Green' 30
bracteosa **32**
brevifolia **34**
Broad-leaved Paperbark **238**, **310**
Broombush **300**
bromelioides **36**
brophyi **38**
caeca 212
cajuputi
    subsp. *cajuputi* **40**
    subsp. *cumingiana* 40
    subsp. *platyphylla* 40
calothamnoides **42**
calycina **44**
    subsp. *dempta* 44
calyptroides 138
campanae **46**
camptoclada 90
capitata 11, **48**
cardiophylla **50**
carrii 214
cheelii **52**
Chenille Honey-myrtle **140**
ciliosa **54**
citrina **56**
citrolens 14
clarksonii 40
clavifolia 292
Claw Honey-myrtle **230**
cliffortioides 220
coccinea 8, **58**
concinna **60**

concreta 6, 11, **62**, 318
condylosa 38
conothamnoides 11, **64**, 304
cordata 11, **66**
Cork-bark Honey-myrtle **276**
cornucopiae **68**
coronicarpa **70**
Cross-leaved Honey-myrtle **84**
croxfordiae **72**
ctenoides **74**
cucullata **76**
cuticularis 8, **78**
dealbata **80**
deanei 256
decora **82**
decussata **84**, 122
delta 70
dempta 44
densa **86**
densispicata **88**
depauperata **90**
depressa **92**, 168
Desert Honey-myrtle **128**
diosmatifolia **94**
diosmifolia 9, **96**, 246
dissitiflora 9, **98**
eleuterostachya **100**
elliptica 8, 11, **102**
ericifolia **104**
erubescens 94
eulobata 46
eurystoma 160
eximia 58, **106**

fabri **108**
filifolia **110**
fissurata 160
fluviatilis 22
foliolosa 192
fulgens 9, 240, 242
  subsp. *corrugata* **112**
  subsp. *fulgens* 112, **114**
  subsp. *steedmanii* **116**
  'orange-flowered form' **118**
  'Georgiana Molloy' **120**, 282
gibbosa 84, **122**
glaberrima **124**
glena 200
globifera **126**
glomerata **128**
Goldfields Bottlebrush **58**
Graceful Honey-myrtle **240**
Granite Bottlebrush 102
Granite Honey-myrtle **102**
Grey Honey-myrtle **146**
grieveana 38
groveana **130**
halmaturorum 8, **132**
halophila 284
hamata **134**
hamulosa 24, **136**
haplantha 78
Hillock Bush **144**
hnatiukii 126
hollidayi **138**
holosericea 250
howeana 6
huegelii
  subsp. *huegelii* 76, **140**
  subsp. *pristicensis* 140
huttensis **142**
hypericifolia 8, **144**
idana 250
incana
  subsp. *incana* **146**
  subsp. *tenella* **148**
  'Velvet Cushion' 146
Inland Paperbark **128**
irbyana **150**, 208
johnsonii 284
keigheryi 304
kunzeoides 130
laetifica 9, **152**
lanceolata **154**, 314
lara 54
lasiandra **156**
lateralis **158**
lateriflora
  subsp. *acutifolia* 160
  subsp. *lateriflora* **160**
lateritia 8, **162**
laxiflora 74, **164**
lecanantha 158
leiocarpa **166**
leiopyxis 8, **168**
leptospermoides 248, 298
leucadendra 8, 11, **170**
leuropoma **172**
linariifolia 8, 9, 16, **174**
  'Snowstorm' 174
linguiformis 294
linophylla 11, **176**
longistaminea
  subsp. *longistaminea* 9, 178
  subsp. *spectabilis* **178**
macronychia 8,
  subsp. *macronychia* **180**
  subsp. *trygonoides* 180
Mallee Honey-myrtle **34**
manglesii **182**, 252
megacephala 11, **184**
micromera **186**
microphylla 8, **188**, 190
  aff. *microphylla* **190**
minutifolia 11, **192**
Moonah **154**
monantha 192
nanophylla 192
Narrow-leaved Paperbark **174**
nematophylla **194**, 196
  aff. *nematophylla* **196**
nervosa 9
  subsp. *nervosa* **198**
  subsp. *crosslandiana* 198
nesophila 8, **200**
  'Little Nessie' 200
nodosa **202**
oldfieldii **204**
osullivanii 6, **134**
orbicularis 66
ordinifolia 34
oxyphylla **206**
pallescens **208**
papillosa 252
parviceps 182
parvistaminea 104
pauciflora 222
pauperiflora
  subsp. *fastigiata* 210
  subsp. *mutica* **210**
  subsp. *pauperiflora* 210
penicula 58, 106
pentagona 9,

var. *latifolia* 8, **212**
var. *pentagona* **214**
var. *raggedensis* **214**
phoidophylla 20
platycalyx **216**
plumea 11, 60, **218**
podiocarpa **220**
polycephala **222**
pomphostoma 32
Prickly Paperbark, **274**
preissiana **224**, 272
pritzelii 86
procera **226**
psammophila **228**, 316
pulchella **230**
pungens **232**
pustulata **234**
quadrifaria **236**
quinquenervia **238**, 310
radula **240**, 242
 '*radula* hybrid' **242**
rhaphiophylla **244**
rigidifolia 218
ringens 96, **246**
Robin Redbreast Bush **162**
Rough Honey-myrtle **252**
ryeae **248**
saligna 80
Salterwater Paperbark **78**
sapientes **250**
scabra **252**
Scarlet Honey-myrtle **114**
Scented Paperbark **266**
sciotostyla 78
sclerophylla **254**

sculponeata 18
seriata 182
sericea 80
sheathiana 210
Showy Honey-myrtle **200**
sieberi **256**
Silver Cajuput **22**
Silver-leaved Paperbark **22**
similis 268
societatis **258**
South Australian Swamp Paperbark **132**
sparsiflora 132
spathulata 11, **260**
spicigera **262**
squamea **264**
squamophloia 274
squarrosa 11, **266**
stenostachya 80
stereophloia 300
stipitata 68
stramentosa 268
striata 9, 11, **270**
strobophylla **272**
styphelioides 8, **274**
suberosa **276**
subfalcata **278**
subtrigona 258
subularis 132
Swamp Honey-myrtle **264**
Swamp Paperbark **104**, **244**
systena **280**
tamariscina 150, 208, 290
teretifolia 11, 120, **282**
teuthidoides **294**

thapsina 9, **284**
thymifolia **286**
thymoides **288**
thyoides **290**
tinkeri **292**
torquata **294**
tortifolia 52
Totem Poles **84**
trichophylla 252, 292, **296**
trichostachya 174
triumphalis 198
tuberculata **298**
 var. *arenaria* 298
 var. *macrophylla* 298
 var. *tuberculata* 298
uncinata 6, 62, **300**
undulata 70
un-named species 1 **316**
un-named species 2 **318**
urceolaris **302**
venusta **304**
villosisepala 226
viminea 9,
 subsp. *appressa* 306
 subsp. *demissa* 306
 subsp. *viminea* **306**
violacea **308**
virgata 302
viridiflora **310**
Weeping Paperbark **170**
wilsonii **312**
Wilson's Honey-myrtle **312**
Wiry Honey-myrtle **194**
wonganensis 226
xerophila **314**
zonalis 92